Studies in Systems, Decision and Control

Volume 6

Series editor

Janusz Kacprzyk, Polish Academy of Sciences, Warsaw, Poland
e-mail: kacprzyk@ibspan.waw.pl

For further volumes:
http://www.springer.com/series/13304

About this Series

The series "Studies in Systems, Decision and Control" (SSDC) covers both new developments and advances, as well as the state of the art, in the various areas of broadly perceived systems, decision making and control- quickly, up to date and with a high quality. The intent is to cover the theory, applications, and perspectives on the state of the art and future developments relevant to systems, decision making, control, complex processes and related areas, as embedded in the fields of engineering, computer science, physics, economics, social and life sciences, as well as the paradigms and methodologies behind them. The series contains monographs, textbooks, lecture notes and edited volumes in systems, decision making and control spanning the areas of Cyber-Physical Systems, Autonomous Systems, Sensor Networks, Control Systems, Energy Systems, Automotive Systems, Biological Systems, Vehicular Networking and Connected Vehicles, Aerospace Systems, Automation, Manufacturing, Smart Grids, Nonlinear Systems, Power Systems, Robotics, Social Systems, Economic Systems and other. Of particular value to both the contributors and the readership are the short publication timeframe and the world-wide distribution and exposure which enable both a wide and rapid dissemination of research output.

Ramón González · Francisco Rodríguez
José Luis Guzmán

Autonomous Tracked Robots in Planar Off-Road Conditions

Modelling, Localization, and Motion Control

Springer

Ramón González
Department of Computer Science
University of Almeria
Ctra. Sacramento s/n
04120 Almeria
Spain
E-mail: rgonzalez@ual.es

José Luis Guzmán
Department of Computer Science
University of Almeria
Ctra. Sacramento s/n
04120 Almeria
Spain

Francisco Rodríguez
Department of Computer Science
University of Almeria
Ctra. Sacramento s/n
04120 Almeria
Spain

Additional material to this book can be downloaded from http://extras.springer.com

ISSN 2198-4182 ISSN 2198-4190 (electronic)
ISBN 978-3-319-38332-3 ISBN 978-3-319-06038-5 (eBook)
DOI 10.1007/978-3-319-06038-5
Springer Cham Heidelberg New York Dordrecht London

Printed on acid-free paper

Springer is part of Springer Science+Business Media (www.springer.com)

You have to test your stuff a lot,
test it just like you're going to fly it,
and then fly it just like you tested it.

Professor Steven W. Squyres
Principal Investigator
Mars Exploration Rover mission

Contents

Chapter 1
Introduction

This chapter addresses a historical perspective of tracked vehicles and motivates the use of tracked mobile robots. It highlights the main contributions of this monograph. Finally, the main assumptions and limitations regarding the developed work are explained.

1.1 The Importance of Tracked Vehicles

1.1.1 Tracked Vehicles along History

The origins of tracked vehicles are closely motivated by the idea of developing cross-country vehicles, that is, vehicles moving on off-road rough terrains (packed snow, muddy roads, loose sandy soils, etc.). In 1770, a patent by Richard L. Edgeworth firstly introduced the concept of what is known today as full-track [6]: "the invention cossets in making portable railways to wheel carriages so that several pieces of wood are connected to the carriage which it moves in regular succession in such a manner that a sufficient length of railing is constantly at rest for the wheels, to roll upon, and that when the wheels have nearly approached the extremity of this part of the railway, their motions shall lay down a fresh length of rail in front, the weight of which in its descent shall assist in raising such part of the rail as the weeks have already passed over, and thus the pieces of wood which are taken up in the rear are in succession laid in the front, so as to furnish constantly a railway for the weeks to roll on." By those years none of the inventions became successful products. A further step came with the introduction of internal-combustion engines at the end of the 19th century. In this sense, a clear contribution was the Holt tractor in 1911 (Figure 1.1a). This vehicle was widely used by the British, French, and American armies in World War I for hauling heavy equipment. This successful platform led to the development of war tanks.

After the War, the prolific research carried out was applied to agriculture. Indeed during the years 1920-1940, many fundamental questions were unfolded such as: the relationship between track and soil, the movement re-

R. González, F. Rodríguez, and J.L. Guzmán, *Autonomous Tracked Robots in Planar Off-Road Conditions,* Studies in Systems, Decision and Control 6,
DOI: 10.1007/978-3-319-06038-5_1, © Springer International Publishing Switzerland 2014

sistance, sinkage, wheel dimensions, "ground pressure", etc. That research led to consider the development of tracked vehicles as a solid discipline [6].

At this point, it is interesting to introduce a brief discussion comparing the performance of wheels and tracks. In this regard the article [133] constitutes a solid reference. There, after several comparisons on different terrains (sand, clay, loam), authors demonstrate that the thrust developed by a wheeled vehicle will generally be lower than that developed by a comparable tracked vehicle, this is particularly noticeable on cohesive soils. This is motivated by the fact that the average normal pressure under the tires of a wheeled vehicle will be higher than that under a comparable tracked vehicle. The paper [54] states that the wheeled-vs-track dilemma is rooted in the following fundamental subjects: the vehicle's mission, the terrain profile, and the specific vehicular features. In particular, the main advantages of tracked vehicles are:

- They offer the best solution for a versatile platform that is required to operate over diverse terrains even at different weather conditions. For that reason this kind of vehicles are very common in search and rescue operations especially on snow.
- They generate low ground pressure, which leads to conserve the natural environment. This motivates why tracked vehicles are employed in agriculture (harvesting, planting, or spraying tasks) and mining activities (Figure 1.1b, d).
- They prevent from sinking, even becoming stuck, into soft ground, therefore they are ideal vehicles for loose sandy terrains like in military operations (hauling heavy artillery or as personnel transportation units) (Figure 1.1a, c).

1.1.2 Autonomous Tracked Vehicles

It is clear that tracked vehicles comprise a solid and successful way of transportation, especially in off-road conditions. For that reason, when a roboticist thinks about a mobile robot to operate in off-road conditions one of the first options is to use tracked locomotion. However the task of making autonomous a (tracked) vehicle leads to consider many challenging issues.

The behaviour of a mobile robot in outdoor off-road conditions is quite different from a mobile robot working on structured indoor environments. The most important disturbances and inconveniences affecting to off-road mobile robots deal with the robot-terrain interactions, such as slip and sinkage phenomena. The first issue to be solved is related to the robot mechanics. Off-road applications generally require robots to travel across unprepared rough terrains. This fact means that proper locomotion mechanisms and on-board sensors should be selected to safely steer the robot autonomously [53, 117]. Furthermore, due to the robot can work in remote or inaccessible scenarios, efficient power units and reliable communication systems are also required. Figure 1.2 shows several tracked mobile robots applied in challenging off-road

(a) Holt tractor, applied to hauling heavy artillery during World War I (year 1911) [6]

(b) Komatsu bulldozer D50, applied to construction and mining (year 1947) [97]

(c) Hagglunds Bv 206 vehicle, applied to military personnel transportation (year 1974) [52]

(d) Caterpillar D6M, applied to agriculture (year 1986) (Source www.cat.com)

Fig. 1.1 Some of the most successful vehicles and companies dealing with tracked locomotion

conditions. In particular, Figure 1.2a deals with the robot *Fitorobot* developed for working inside greenhouses mainly in spraying activities [43]; Figure 1.2b presents the robot Auriga-β for fire extinction tasks [98]; next figure is related to a research platform developed at the University of Minnesota in which the tracked mechanical configuration is adapted in order to climb stairs [100]; Figure 1.2d shows the well-known Packbot robot from iRobot, this tracked robot is employed by the US troops to perform activities such as: explosives detection and surveillance/reconnaissance, among others.

(a) Fitorobot (University of Almeria, Spain), agriculture (spraying activities) [43]

(b) Auriga-β (University of Malaga, Spain), fire extinction tasks [98]

(c) Research platform (University of Minnesota, USA), stair climbing robot [100]

(d) Packbot (iRobot, USA), military activities

Fig. 1.2 Some examples of successful tracked mobile robots applied in off-road conditions

As previously introduced, one fundamental factor that can affect the performance of the motion of off-road mobile robots is the slip phenomenon. For these reasons, the design of trajectory planners, motion controllers and localization strategies, taking into account this unavoidable effect, constitutes a key issue in the field of off-road mobile robotics. For instance, in the work [46], the kinematic modelling of a mobile robot, with and without slip, considering several types of wheels is examined. These models are obtained from physical principles. In [129], a study for four generic Wheeled Mobile Robots (WMR) in the presence of wheel skidding and slip from a control perspective is developed.

Perturbations due to skidding and slip are categorically classified as input-additive, input-multiplicative, and/or matched/unmatched perturbations. In [125], an interesting software tool is presented evaluating the performance of different robot locomotion mechanisms in off-road rough environments. The works [33] and [78] address the problem of slip for an Ackermann-type agricultural vehicle in which adaptive and predictive control techniques were used to face the lateral slip effects. The work [76] contributes a comprehensive dynamic model of a tracked mobile robot that can be used to estimate key soil properties upon which the robot operates. For that purpose, track slips, the vehicle slip angle, and track forces are considered.

This chapter is organized as follows. Section 1.2 motivates and highlights the main contributions of this monograph. An outline is addressed in Section 1.3. Finally, the main assumptions are stated in Section 1.4.

1.2 Motivation and Contributions

When a mobile robot moves in off-road conditions, some undesirable phenomena, such as noisy measurements, slip and sinkage effects, changing light conditions, and vibrations, among others; lead to inaccurate robot location and influence negatively the controllability of the robot. Furthermore, the design of motion controllers usually requires simplified models and, sometimes, some dynamics are too complex to be expressed by a set of mathematical equations. For these reasons, modelling, localization and control issues related to off-road mobile robots are challenging research topics for the robotics and automatic control communities.

From the point of view of the automatic control community, there are many settled control strategies that can deal with such constraints and uncertainties. However, motion control strategies that comprise both phenomena are rarely found in the off-road mobile robotics literature. One of the main reasons for this fact is that elaborate control strategies need a considerable computation time, what in off-road mobile robot applications is really undesirable due to the low performance of the computing resources and the low sampling periods.

On the other hand, in order to achieve an appropriate control of the off-road mobile robot, reliable "feedback" must be obtained from the robot-environment interaction, particularly the robot location. In this case, localization techniques have to deal with the special features of off-road conditions (previously depicted). Generally, popular solutions such as wheel-based odometry and dead-reckoning could become unsuitable. Even sometimes, GPS-based solutions are not applicable, for instance, in space exploration or in partially covered environments. To overcome these shortcomings, probabilistic techniques, such as Kalman filter or Particle filter, are usually considered combining data from different sources. The main drawback of these approaches are that several (redundant) sensors must be available. One of the most successful approaches to localize mobile robots in off-road environments is called visual odometry. This method estimates the robot motion (location) using a sequence of images from an on-robot camera(s). One of the most challenging issues about this technique deals with the "error growth" phenomenon along long trajectories.

This monograph presents some state of the art and several novel solutions of the specific issue of tracked mobile robots working in planar off-road conditions. In this sense, this monograph focuses on obtaining appropriate models and localization techniques for tracked mobile robots where the slip phenomenon has a high influence (Chapters 2 and 3). Then, these models and localization methods are used to design different motion control strategies for a proper robot navigation in spite of the slip effects (Chapters 4 and 5).

First, an extended kinematic model for a tracked mobile robot (differential-drive mechanism) taking into account slip effect has been suggested. This model avoids the estimation of complex variables usually related to dynamic models. Afterwards, a kinematic model for the trajectory tracking problem is obtained. Different formulations of that trajectory tracking error model are suggested (continuous-time, discrete-time, and discrete-time with additive uncertainty), which will be employed to design the motion controllers subsequently. Physical experiments show that the kinematic model extended with slip is more accurate than the classic kinematic one when a mobile robot moves in off-road conditions.

Regarding the localization issue, a visual-odometry-based technique is suggested. The most interesting point is that two cameras are combined, one for estimate the robot longitudinal displacement and another camera to estimate the robot orientation (visual compass). In this way, typical problems related to error growth of odometry-based solutions and false-matching phenomena of visual-odometry-based approaches are minimized.

Once the robot models and the localization techniques are developed and described, three motion control strategies are suggested. First, the well-known linear feedback controller [21, 64] is modified considering slip effect (Chapter 4). In this case, time-dependent feedback gains are updated online using the estimated slip. The main limitation of this control strategy is that state and input constraints fulfillment and stability are not ensured. In order to overcome these limitations, more elaborate control strategies are presented. On the one hand, an adaptive control scheme formulated using Linear Matrix Inequalities (LMI) is proposed (Chapter 4). The main advantages of this control strategy are: asymptotic stability, performance, and input and state constraints fulfillment. They have been guaranteed through the determination of an ellipsoidal invariant set and a quadratic Lyapunov function. Efficient real-time execution has also been achieved solving online an LP problem that produces the current feedback control gain as a convex combination of a set of offline gains (one gain for each extreme realization of the system dynamics). The main limitation of this control scheme is that uncertainties are not considered. For that reason, a robust Model-based Predictive

Control (MPC) strategy has been proposed (Chapter 5). Particularly, this controller has been formulated taking advantage of the concept of "tubes". The main advantages of this control strategy are: robustness (uncertainties are taken into account in the controller synthesis), performance, input and state constraints fulfillment, stability, and efficient real-time execution.

Notice that several robust control techniques exist in the literature, such as Quantitative Feedback Theory (QFT) [56], control based on H_∞-norm [138], and μ-synthesis control [107], among others. For a summary see [4] and the references therein. In this monograph, the robust MPC approach has been selected due to the following goodness: an optimization problem of the type Quadratic Programming (QP) is solved at each sampling instant obtaining the proper control actions as a compromise between small deviations from the reference trajectory and suitable control actions; it handles state and input constraints directly in the optimization problem; it easily handles multi-dimensional variables; easy tuning; reference trajectory is known *a priori*; the model related to robot motion is known, what means that future evolution of robot motion can be predicted; there are several settled ways to robustify standard MPC controllers (Open-loop Min-Max MPC, Feedback Min-Max MPC, H_∞ MPC, Tube-based MPC, etc.). For a detailed description about robust MPC see [8, 18, 96, 111], and the references therein.

Finally, in order to validate previous contributions through physical experiments, a mobile robot has been employed (see Figure 1.2a) [43]. The mobile robot has a mass of 500 [Kg] and its dimensions are 1.5 [m] long × 0.7 [m] wide. It is driven by a 20 [HP] gasoline engine.

In relation to physical experiments, a typical four-layer navigation architecture has been developed, which will run on the computer on-board the mobile robot *Fitorobot* (Figure 1.3).

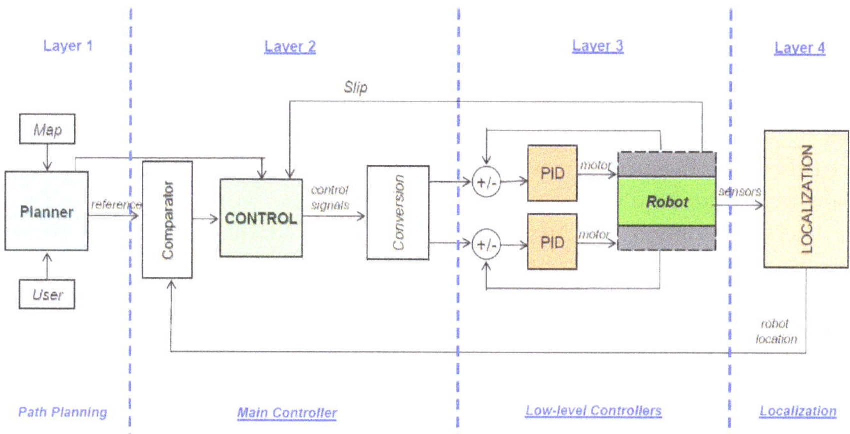

Fig. 1.3 Navigation architecture

The first layer is devoted to path planning. This layer generates the reference or desired trajectories and velocities. In this monograph, it is considered that the reference trajectory has been defined before the experiment starts. The second layer includes the motion controllers previously introduced. In the third layer, two low-level controllers have been implemented in order to ensure that the set-points generated by the motion controllers are reached by the robot tracks. For that purpose, two PI controllers with an anti-windup scheme have been tuned (see [42] for details). The fourth layer is related to the robot localization. In this case, the vision-based approach is employed.

1.3 Outline of this Monograph

This monograph is composed of six chapters. This current chapter serves as an introduction and summarizes the main contributions.

In Chapter 2, the trajectory tracking error model using an extended kinematic model is formulated. The kinematic model is extended considering slip. The trajectory tracking error model is discretized. The main purpose of this model is to be employed in the design of slip compensation control strategies for off-road mobile robots.

Chapter 3 deals with the vision-based approach.

In Chapter 4, adaptive slip compensation controllers are presented. First, a linear feedback controller is formulated in a way that slip is used to modify the feedback gains. Afterwards, an adaptive control strategy is formulated using LMI.

Chapter 5 focuses on a robust control strategy. In this sense, a robust predictive controller have been designed such that system uncertainties and constraints are taken into account in the controller synthesis and efficient computation time is also achieved.

Chapter 6 is devoted to conclusions and future research.

1.4 Assumptions and Limitations

At this point, five primary assumptions related to the work presented in this monograph are highlighted.

The first assumption, and which is remarked in the title of this monograph, is that the tracked mobile robots considered move mainly in planar off-road conditions. This assumption avoids the problem of stable motion of tracked vehicles on bumpy or stairs-like profiles (Figure 1.4). It is considered that this problem is closer to a mechanical solution rather than the control perspective addressed in this monograph. As deep discussion about this issue is introduced in [47]. Anyway, high obstacles cannot be negotiated due to the inability of tracked vehicles to change the track shape configuration or its payload position. Even the limited suspension system.

Fig. 1.4 Tracked locomotion on bumpy (or 3D) terrains [47]

The second assumption is that the mobile robots under consideration move at slow velocities (≤ 2 [m/s]). Thus, lateral slip and forces arising from dynamic effects are neglected. As stated in [66], [76], [78], lateral slip is zero for straight-line motions and it can be neglected when the vehicle turns "on the spot" or at low velocities. However, longitudinal slip is an unavoidable effect of the robot-terrain interaction [129], [132].

The third assumption is that the robot tracks are stiff relative to the terrain, that is, the case of a track moving over a deformable terrain is considered [58].

The fourth assumption is that the robot is moving on narrow workspaces. It means if the error between the reference trajectory and the real robot trajectory is too large, the robot can crash with obstacles. The maximum allowed deviation will be constrained according to each particular scenario.

The last assumption is that the environment, in which the mobile robot is moving, is static and traversable.

Chapter 2
Modelling Tracked Robots in Planar Off-Road Conditions

This chapter focuses on kinematic modelling of tracked mobile robots in off-road slippery conditions. In this sense, a trajectory tracking error model based on an extended kinematic model considering slip effect is formulated. This model avoids the estimation of complex variables usually related to dynamic models. Physical experiments show that the extended kinematic model is more accurate than the classic kinematic model when a tracked robot moves in off-road conditions.

2.1 Introduction

In the context of robotics, a model is defined as a set of mathematical differential equations that represents the behaviour of a robot. In this sense, kinematic and dynamic modelling constitutes a key issue related to off-road mobile robotics, since such models may be used to design appropriate mechanical structures [117, 125]; to design planning algorithms and motion controllers [58]; to estimate the robot localization [123]; and to implement software simulators [48], among others.

In this book, the modelling issues are mainly related to the design of proper control strategies compensating/minimizing the slip phenomenon [50, 58, 61, 71, 79, 137]. Therefore, this chapter focuses on obtaining a model of a Tracked Mobile Robot (TMR) moving on slip conditions to be used for control purposes. Notice that there is a vast source of papers dealing with the performance of tracked vehicles, there comprehensive models and detailed analyses from a terramechanical point of view are found [6, 9, 36, 59, 66, 81, 108, 132]. In this monograph such models have been avoided since they are quite complicated even they require numerous parameters some of them very difficult to measure or to estimate.

As previously introduced, when a mobile robot operates in off-road conditions, modelling problem requires to study and to analyze several phenomena that affect to the robot mobility and controllability. In this sense, slip constitutes one of the most important phenomenon dealing with the robot-terrain

R. González, F. Rodríguez, and J.L. Guzmán, *Autonomous Tracked Robots in Planar Off-Road Conditions,* Studies in Systems, Decision and Control 6,
DOI: 10.1007/978-3-319-06038-5_2, © Springer International Publishing Switzerland 2014

interaction. Slip is defined as the difference between the velocity measured by a track (theoretical velocity) and the actual forward vehicle velocity (real linear velocity) [132]. Mathematically

$$i_j = 1 - \frac{v}{\rho \omega_j}, \tag{2.1}$$

where i_j is the slip of track j, v is the forward vehicle velocity, ρ is the track radius (sprocket) and ω_j is the angular velocity of track j. Physically, it means that, if the vehicle is subject to slip ($v < \rho\omega_j$), the distance that it should travel is shorter than that which the vehicle would travel in an ideal case where no slip is experienced[1] (see Figure 2.1a).

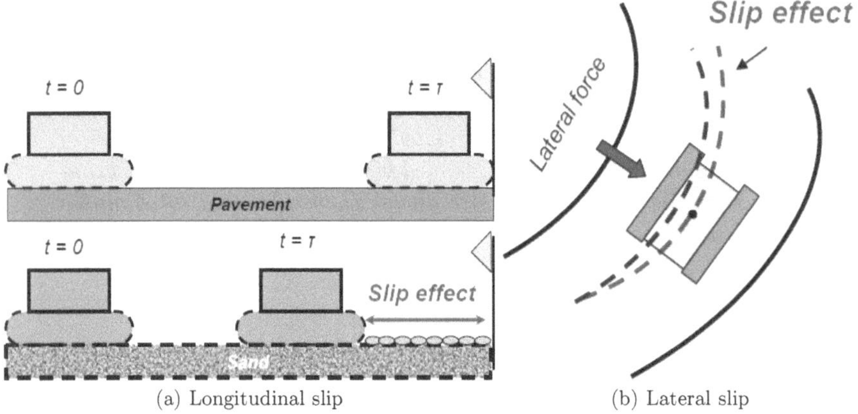

(a) Longitudinal slip (b) Lateral slip

Fig. 2.1 Physical meaning of slip phenomena

In the case of tracked vehicles (or vehicles with rigid wheels), the slip phenomenon is caused by different reasons. Lateral slip is only shown in vehicles that turn to high velocities, and it is only related to the centrifugal force. For that reason, when the tracked vehicle turns on the spot or moves at middle/low velocities, it can be neglected [1, 66, 76]. Longitudinal slip is due to more complex processes. The most important process that motivates longitudinal slip is the track-soil interaction. This effect deals with the properties of the soil under the vehicle (deformation, cohesion, etc.), and the own vehicle parameters (weight, track width, etc.). Another issue is the vertical stress or sinkage that causes the vehicle sinks in the soil (especially in sand). This produces a bulldozing effect along the motion of the vehicle [66, 76, 132],

[1] A special case of slip is when there is a locked track, in which the angular velocity is zero, whereas the linear velocity is not zero. Under this condition, slip can be negative ($v > r\omega_j$). In this book, this particular situation is not considered since, in the employed testbed, the tracks never become locked. For a detailed discussion see [62, 132].

that implicitly influences the horizontal stress or longitudinal slip (see Figure 2.1a). Slip also depends on the track velocity, although for relatively small velocities (as this book case $\leq 2[m/s]$), velocity is not a significant factor, compared to the terrain properties [2].

The main factors causing slip phenomenon in tracked vehicles are summarized in the following:

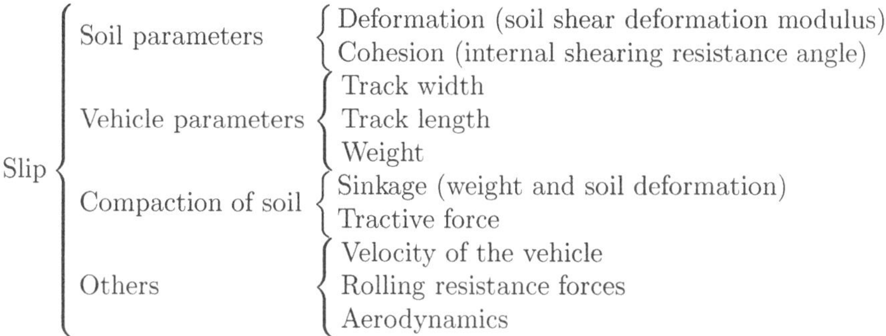

As previously commented, this chapter focuses on obtaining a model of a TMR moving on slip conditions. To this end, the Classic Kinematic Model (CKM) of a differential-drive mobile robot [20], [118] has been extended to include the slip effect and resulting into the Extended Kinematic Model (EKM). As found in the literature, for instance in [88], [98], the difference between WMR (unicycle model) and TMR is that whereas the Instantaneous Centres of Rotation (ICR) for Wheeled Mobile Robots (WMR) are constant and coincident with the ground contact points of the wheels, the ICR for TMR are variable and lie outside the track centrelines due to lateral slip. However, as assumed in this book (see Section 1.4), no lateral slip is considered, what means that the kinematic model of a TMR can be approximated by the model of a differential-drive WMR, that is, we are considering the known unicycle model but taking into account slip phenomenon. Notice that similar kinematic models are employed in numerous works, see for instance, [31], [32], [77], [128], [136], [137]. Such extended kinematic model is used for formulating a new trajectory tracking error model where slip will play a major role. This model will be discretized and an additive term will be included taking into account uncertainty. The objective of these models will be the synthesis of several motion control strategies proposed in Chapters 4 and 5.

This chapter is organized as follows. Section 2.2 presents the extended kinematic model. In Section 2.3, the trajectory tracking error model is detailed. In Section 2.4, the trajectory tracking error model is discretized and extended considering uncertainty. Section 2.5 deals with state and input constraints. In Section 2.6, different techniques to estimate slip in real-time are reviewed and the two selected strategies are explained. In Section 2.7, experimental results validating the extended kinematic model are discussed and the uncertainty term is analytically derived. Finally, Section 2.8 is devoted to conclusions and future research.

2.2 Extended Kinematic Model with Slip

As known, in absence of track slip, the linear velocity of the tracks is given by

$$\begin{aligned} v_r(t) &= \rho\omega_r(t), \\ v_l(t) &= \rho\omega_l(t), \end{aligned} \tag{2.2}$$

where $v_r \in \mathbb{R}$ and $v_l \in \mathbb{R}$ are the linear velocities of the right and left tracks, respectively, $t \in \mathbb{R}^+$ is the continuous time, and $\omega_r \in \mathbb{R}$ and $\omega_l \in \mathbb{R}$ are the angular velocities of the right and left tracks, respectively. In this way, the kinematic model for a differential-drive mobile robot is given by [20], [118]

$$\begin{aligned} \dot{x}(t) &= \frac{v_r(t) + v_l(t)}{2}\cos\theta(t), \\ \dot{y}(t) &= \frac{v_r(t) + v_l(t)}{2}\sin\theta(t), \\ \dot{\theta}(t) &= \frac{v_r(t) - v_l(t)}{b}, \end{aligned} \tag{2.3}$$

where $[x \quad y \quad \theta]^T \in \mathbb{R}^3$ represents the location (position and orientation) of the mobile robot and $b \in \mathbb{R}^+$ is the distance between tracks centres[2] (see Figure 2.2).

However, when the robot moves on slip conditions, the real velocity of the tracks will be lower than the theoretical velocity of the tracks by a penalizing factor [132]

$$\begin{aligned} v_r^{sl}(t) &= v_r(t)\left(1 - i_r(t)\right) = \rho\omega_r(t)\left(1 - i_r(t)\right), \\ v_l^{sl}(t) &= v_l(t)\left(1 - i_l(t)\right) = \rho\omega_l(t)\left(1 - i_l(t)\right), \end{aligned} \tag{2.4}$$

where $v_r^{sl} \in \mathbb{R}$ and $v_l^{sl} \in \mathbb{R}$ are the linear velocities under slip effect ($v_r^{sl} \leq v_r$, $v_l^{sl} \leq v_l$), and $i_r \in \mathbb{R}^+$ and $i_l \in \mathbb{R}^+$ are the right and left track slip estimations, respectively.

[2] Notice that given a point O_L centered between the two drive wheels, each wheel is a distance $b/2$ from O_L; forward spin of right wheel (v_r) results in counter-clockwise rotation at point O_L. The same calculation applies to the left wheel (v_l), with the exception that forward spin results in clockwise rotation at point O_L [118].

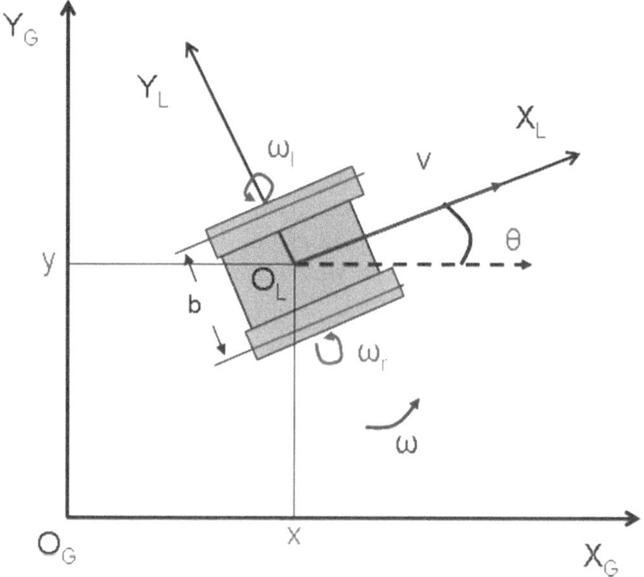

Fig. 2.2 Motion of a tracked mobile robot ($O_G X_G Y_G$ is the reference or inertial frame and $O_L X_L Y_L$ is the local robot frame)

In this sense, taking into account the classic formulation of the kinematic model for a differential-drive mobile robot (2.3), the continuous-time kinematic model of a robot on slip conditions is given by [2], [77], [136], [137]

$$\begin{aligned} \dot{x}^{sl}(t) &= \frac{v_r^{sl}(t) + v_l^{sl}(t)}{2} \cos\theta^{sl}(t), \\ \dot{y}^{sl}(t) &= \frac{v_r^{sl}(t) + v_l^{sl}(t)}{2} \sin\theta^{sl}(t), \\ \dot{\theta}^{sl}(t) &= \frac{v_r^{sl}(t) - v_l^{sl}(t)}{b}. \end{aligned} \tag{2.5}$$

where $[x^{sl} \quad y^{sl} \quad \theta^{sl}]^T \in \mathbb{R}^3$ represents the location of the real mobile robot on slip conditions.

For notational convenience, the model can also be written as

$$\begin{aligned} \dot{x}^{sl}(t) &= v^{sl}(t)\cos\theta^{sl}(t), \\ \dot{y}^{sl}(t) &= v^{sl}(t)\sin\theta^{sl}(t), \\ \dot{\theta}^{sl}(t) &= \omega^{sl}(t), \end{aligned} \tag{2.6}$$

where $v^{sl} = (v_r^{sl} + v_l^{sl})/2$ is the forward robot velocity and $\omega^{sl} = (v_r^{sl} - v_l^{sl})/b$ is the angular robot velocity.

2.3 Extended Trajectory Tracking Error Model with Slip

Trajectory tracking consists in the problem in which a robot must follow a *reference or virtual* mobile robot representing the desired locations and velocities (see Figure 2.3). Notice that the trajectory tracking problem is related to control the location and velocity of the robot [21], [64], while path tracking only considers the location [22], [106]. In the latter case, the velocity is viewed as a measured parameter, which is manually controlled.

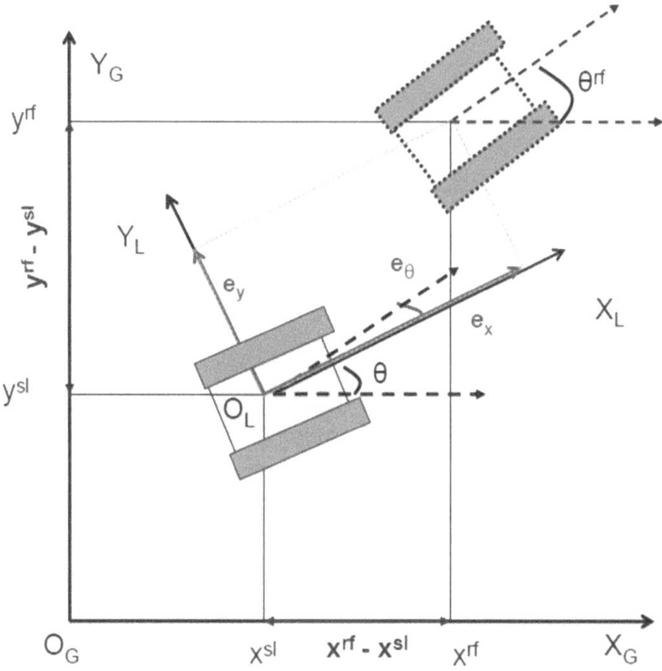

Fig. 2.3 Graphical representation of the trajectory tracking problem. The objective is to reduce the error between the *reference or virtual* robot (dotted lines) and the real robot (solid lines)

The objective of the trajectory tracking problem is to find a control law, such that the error between the desired and the actual location of the mobile robot is close to zero at any given time (regulation problem). Thus, a trajectory tracking error model using the previous EKM (2.5), (2.6) is presented in this section to be used for that purpose.

Suppose that the reference robot follows a trajectory calculated through unicycle kinematics, that is,

$$\begin{aligned}\dot{x}^{rf}(t) &= \frac{v_r^{rf}(t) + v_l^{rf}(t)}{2}\cos\theta^{rf}(t),\\ \dot{y}^{rf}(t) &= \frac{v_r^{rf}(t) + v_l^{rf}(t)}{2}\sin\theta^{rf}(t),\\ \dot{\theta}^{rf}(t) &= \frac{v_r^{rf}(t) - v_l^{rf}(t)}{b},\end{aligned} \tag{2.7}$$

where $[x^{rf} \quad y^{rf} \quad \theta^{rf}]^T \in \mathbb{R}^3$ represents the location of the reference robot and $v_r^{rf} \in \mathbb{R}, v_l^{rf} \in \mathbb{R}$ are the desired velocities for the right and left tracks, respectively.

For notational convenience, the model is also written as

$$\begin{aligned}\dot{x}^{rf}(t) &= v^{rf}(t)\cos\theta^{rf}(t),\\ \dot{y}^{rf}(t) &= v^{rf}(t)\sin\theta^{rf}(t),\\ \dot{\theta}^{rf}(t) &= \omega^{rf}(t),\end{aligned} \tag{2.8}$$

where $v^{rf} = (v_r^{rf} + v_l^{rf})/2$ is the forward reference velocity, and $\omega^{rf} = (v_r^{rf} - v_l^{rf})/b$ is the angular reference velocity.

As discussed in the previous section, the real mobile robot will move on slip conditions, and thereby, it is modelled by (2.5). Then, the error or deviation between the reference and the real robot, expressed in relation to the real robot frame, is given by

$$\begin{bmatrix} e_x(t) \\ e_y(t) \\ e_\theta(t) \end{bmatrix} = \begin{bmatrix} \cos\left(\theta^{sl}(t)\right) & \sin\left(\theta^{sl}(t)\right) & 0 \\ -\sin\left(\theta^{sl}(t)\right) & \cos\left(\theta^{sl}(t)\right) & 0 \\ 0 & 0 & 1 \end{bmatrix} \begin{bmatrix} x^{rf}(t) - x^{sl}(t) \\ y^{rf}(t) - y^{sl}(t) \\ \theta^{rf}(t) - \theta^{sl}(t) \end{bmatrix}, \tag{2.9}$$

where $e_x \in \mathbb{R}$ is the longitudinal error, $e_y \in \mathbb{R}$ is the lateral error and $e_\theta \in \mathbb{R}$ is the orientation error. These errors are graphically represented in Figure 2.3.

Differentiation of (2.9) with respect to time yields

$$\begin{aligned}\begin{bmatrix} \dot{e}_x \\ \dot{e}_y \\ \dot{e}_\theta \end{bmatrix} &= \begin{bmatrix} 0 & \omega^{sl} & 0 \\ -\omega^{sl} & 0 & 0 \\ 0 & 0 & 0 \end{bmatrix} \begin{bmatrix} e_x \\ e_y \\ e_\theta \end{bmatrix} + \begin{bmatrix} \cos e_\theta \\ \sin e_\theta \\ 0 \end{bmatrix} v^{rf} - \\ &- \begin{bmatrix} 1 \\ 0 \\ 0 \end{bmatrix} \frac{v_r + v_l}{2} + \begin{bmatrix} 0 \\ 0 \\ \left(\frac{v_r^{rf} - v_l^{rf}}{b} - \frac{v_r - v_l}{b}\right) \end{bmatrix} + \begin{bmatrix} \frac{v_r i_r + v_l i_l}{2} \\ 0 \\ \frac{v_r i_r - v_l i_l}{b} \end{bmatrix}.\end{aligned} \tag{2.10}$$

Finally, the non-linear model (2.10) can be linearized around the reference trajectory (see Remarks 1 and 2), following two approaches. The first case consists in to define the variable

$$\lambda = \frac{v_r^{rf} i_r - v_l^{rf} i_l}{b}. \tag{2.11}$$

Afterwards, a first-order Taylor expansion is carried out, obtaining

$$\begin{bmatrix} \dot{e}_x \\ \dot{e}_y \\ \dot{e}_\theta \end{bmatrix} = \begin{bmatrix} 0 & \omega^{rf} - \lambda & 0 \\ -(\omega^{rf} - \lambda) & 0 & v^{rf} \\ 0 & 0 & 0 \end{bmatrix} \begin{bmatrix} e_x \\ e_y \\ e_\theta \end{bmatrix} + \\ + \begin{bmatrix} \frac{-1+i_r}{2} & \frac{-1+i_l}{2} \\ 0 & 0 \\ \frac{-1+i_r}{b} & \frac{1-i_l}{b} \end{bmatrix} \begin{bmatrix} v_r \\ v_l \end{bmatrix} + \begin{bmatrix} \frac{1}{2} & \frac{1}{2} \\ 0 & 0 \\ \frac{1}{b} & \frac{-1}{b} \end{bmatrix} \begin{bmatrix} v_r^{rf} \\ v_l^{rf} \end{bmatrix}, \tag{2.12}$$

In this case, an affine model has been obtained. Now, defining the following variable change

$$u_1 = \frac{-1 + i_r}{2} v_r + \frac{-1 + i_l}{2} v_l + \frac{v_r^{rf}}{2} + \frac{v_l^{rf}}{2}, \tag{2.13}$$

$$u_2 = \frac{-1 + i_r}{b} v_r + \frac{1 - i_l}{b} v_l + \frac{v_r^{rf}}{b} - \frac{v_l^{rf}}{b}, \tag{2.14}$$

the affine model (2.12) becomes

$$\begin{bmatrix} \dot{e}_x \\ \dot{e}_y \\ \dot{e}_\theta \end{bmatrix} = \begin{bmatrix} 0 & \omega^{rf} - \lambda & 0 \\ -(\omega^{rf} - \lambda) & 0 & v^{rf} \\ 0 & 0 & 0 \end{bmatrix} \begin{bmatrix} e_x \\ e_y \\ e_\theta \end{bmatrix} + \begin{bmatrix} 1 & 0 \\ 0 & 0 \\ 0 & 1 \end{bmatrix} \begin{bmatrix} u_1 \\ u_2 \end{bmatrix}, \tag{2.15}$$

which is expressed in state space representation as the following continuous-time linear time-varying system (see Remark 3)

$$\dot{e}(t) = A_\gamma(t) e(t) + B u(t), \tag{2.16}$$

where $e = [e_x \quad e_y \quad e_\theta]^T \in \mathbb{R}^3$ is the state vector (position and orientation errors), $u = [u_1 \quad u_2]^T \in \mathbb{R}^2$ is the input vector (see Remark 4). The system matrix A_γ and the input matrix B are defined as

$$A_\gamma(t) = \begin{bmatrix} 0 & \omega^{rf}(t) - \lambda(t) & 0 \\ -(\omega^{rf}(t) - \lambda(t)) & 0 & v^{rf}(t) \\ 0 & 0 & 0 \end{bmatrix}, B = \begin{bmatrix} 1 & 0 \\ 0 & 0 \\ 0 & 1 \end{bmatrix}. \tag{2.17}$$

Assumption 1. *Assume that reference robot track velocities and slip factors are known at each sampling instant, positive and bounded, i.e.,* $v_r^{rf} \in [v_r^{rf,m}, v_r^{rf,M}]$, $v_l^{rf} \in [v_l^{rf,m}, v_l^{rf,M}]$, $i_r \in [i_r^m, i_r^M]$, *and* $i_l \in [i_l^m, i_l^M]$. *Note that the reference robot always moves forward.*

As previously remarked, an alternative formulation can be obtained using the dynamic feedback linearization procedure [21], [106]. In this way, before linearization (2.12), the following parameters are defined

$$\varsigma(t) = \frac{v_r(t)i_r(t) + v_l(t)i_l(t)}{2}, \tag{2.18}$$

$$\vartheta(t) = \frac{v_r(t)i_r(t) - v_l(t)i_l(t)}{b}. \tag{2.19}$$

Then, a similar first-order Taylor expansion to (2.12) is carried out. Afterwards, according to the feedback linearization procedure, the following virtual control signals are considered

$$u_1'(t) = v^{rf}(t)\cos\left(e_\theta(t)\right) - v(t) + \varsigma(t) - \vartheta(t)e_y(t), \tag{2.20}$$

$$u_2'(t) = \omega^{rf}(t) - \omega(t) + \vartheta(t). \tag{2.21}$$

Now, the trajectory tracking error model becomes

$$\dot{e}(t) = A_\gamma'(t)e(t) + Bu'(t), \tag{2.22}$$

where $u' = [u_1' \quad u_2']^T \in \mathbb{R}^2$ is the input vector, the input matrix B is the same than that defined in (2.17), and the system matrix A_γ' is now defined as

$$A_\gamma'(t) = \begin{bmatrix} 0 & \omega^{rf}(t) & 0 \\ -\left(\omega^{rf}(t) - \vartheta(t)\right) & 0 & v^{rf}(t) \\ 0 & 0 & 0 \end{bmatrix}. \tag{2.23}$$

Notice that this model will be employed in Chapter 4, Section 4.2 to design the slip compensation linear feedback controller.

Remark 1. *Notice that the mismatch between the linearized model and the non-linear system grows for values of e_θ far from zero. It will be shown that, in practice, after the transient, e_θ remains close to zero. Then, the problem can be present at the first instants, due to the initial condition. For that reason, it is assumed that, in practice, the real robot and the reference virtual robot start close.*

Remark 2. *It is important to point out that the system has been linearized around the reference, that is, $v_r = v_r^{rf}$, $v_l = v_l^{rf}$, and $e = 0$. Notice that a similar linearization procedure to that found in the following references [21], [64] has been obtained. The result obtained in this book is also acceptable, since slip is not known when reference velocities are defined. However, now, some terms that depend on slip appear in the model. This fact is really advisable, especially to design the slip compensation control strategies proposed in Chapters 4 and 5.*

Remark 3. *From Assumption 1, a time-varying vector of parameters $\gamma = [v_r^{rf}(t) \quad v_l^{rf}(t) \quad i_r(t) \quad i_l(t)]^T \in \mathbb{R}^4$ can be defined, and a bounding set*

$\Gamma \subseteq \mathbb{R}^4$, such that $\gamma(t) \in \Gamma, \forall t \in \mathbb{R}^+$. For any admissible realization of parameter $\gamma \in \Gamma$, a dynamic matrix denoted as $A_\gamma(t)$ is obtained. It follows that $A_\gamma(t) \in \mathcal{A}^\gamma$ where $\mathcal{A}^\gamma$ is a polytope in $\mathbb{R}^{3\times 3}$.

Remark 4. *Note that, according to (2.13)-(2.14), the linear velocities for each track are obtained as*

$$v_r(t) = \frac{v_r^{rf}(t) - u_1(t) - \frac{b}{2}u_2(t)}{1 - i_r(t)}, \tag{2.24}$$

$$v_l(t) = \frac{-v_l^{rf}(t) + u_1(t) - \frac{b}{2}u_2(t)}{-1 + i_l(t)}. \tag{2.25}$$

where $v_r \in [v_r^m, v_r^M]$ and $v_l \in [v_l^m, v_l^M]$. This leads to bounds on the space of u, which are obtained from the constraints on v_r, v_l, i_r, i_l, v_r^{rf}, v_l^{rf} (see Section 2.5).

2.4 Discrete-Time Trajectory Tracking Error Model and Model Uncertainties

In this section, the trajectory tracking error model (2.16) is discretized to be used for the synthesis of the controllers proposed in Chapters 4 and 5. For that purpose, the Euler discretization with sampling period T_s has been employed. In this way, the following discrete-time linear time-varying system is obtained

$$e(k+1) = A_\gamma(k)e(k) + B_d u(k), \tag{2.26}$$

where $k \in \mathbb{Z}^+$ is the discrete sample, and the system and input matrices, A_γ and B_d, are now defined as

$$A_\gamma(k) = \begin{bmatrix} 1 & T_s(\omega^{rf}(k) - \lambda(k)) & 0 \\ -T_s(\omega^{rf}(k) - \lambda(k)) & 1 & T_s v^{rf}(k) \\ 0 & 0 & 1 \end{bmatrix}, \; B_d = \begin{bmatrix} T_s & 0 \\ 0 & 0 \\ 0 & T_s \end{bmatrix} \tag{2.27}$$

Notice that, for notational convenience, A_γ is used both for continuous-time and discrete-time models. The script (k) or (t) will clarify the discrete-time or continuous-time domain, respectively.

The model (2.26) will be employed in Chapter 4 to design an adaptive control strategy.

Finally, the model represented by (2.26) is reformulated considering additive uncertainty

$$e(k+1) = A_v(k)e(k) + B_d u(k) + w(k), \tag{2.28}$$

where w is the additive uncertainty term satisfying $w \in W$ with W a polytope in state space $\mathbb{R}^3$. Notice that the bounding set W includes: the deviation between the non-linear continuous-time model and the linear discrete-time model, the effects of the noise in the slip estimations, and the uncertainty in the estimation of the robot location (see Section 2.7). In a similar way to Remark 3, $\upsilon = [v_r^{rf}(k) \quad v_l^{rf}(k)]^T \in \mathbb{R}^2$ is defined as a time-varying vector such that $\upsilon(k) \in \Upsilon, \forall k \in \mathbb{Z}^+$, where $\Upsilon \subseteq \mathbb{R}^2$ is a polytope. For any admissible realization of parameter $\upsilon \in \Upsilon$, a dynamic matrix denoted as $A_\upsilon(t)$ is determined. It follows that $A_\upsilon(t) \in \mathcal{A}^\upsilon$ where $\mathcal{A}^\upsilon$ is a polytope in $\mathbb{R}^{3\times3}$.

The system matrix A_υ is defined as

$$A_\upsilon(k) = \begin{bmatrix} 1 & T_s\frac{(1-\bar{i}_r)v_r^{rf}(k)-(1-\bar{i}_l)v_l^{rf}(k)}{b} & 0 \\ -T_s\frac{(1-\bar{i}_r)v_r^{rf}(k)-(1-\bar{i}_l)v_l^{rf}(k)}{b} & 1 & T_s v^{rf}(k) \\ 0 & 0 & 1 \end{bmatrix}, \quad (2.29)$$

where $\bar{i}_r$ and $\bar{i}_l$ are the right and left track nominal slips, respectively.

Assumption 2. *Assume that the nominal slip values are known* a priori. *Particularly, they are defined as the mean slip value of a terrain.*

The nominal model is defined as

$$\bar{e}(k+1) = A_\upsilon(k)\bar{e}(k) + B_d g(k), \quad (2.30)$$

where $\bar{e} = [\bar{e}_x \quad \bar{e}_y \quad \bar{e}_\theta]^T \in \mathbb{R}^3$ is the nominal state, and $g \in \mathbb{R}^2$ is the control input for the nominal system.

The models (2.28) and (2.30) will be employed in Chapter 5 to design a robust predictive control strategy. For that reason, the nominal slip is used in the definition of the system matrix A_υ, since the real values of i_r and i_l are unknown for the predictions[3].

2.5 State and Input Constraints

At this point, it is considered that the trajectory tracking error models (2.26) and (2.28) are both subject to constraints. These constraints are imposed on the states and on the control signals. Regarding the state constraints, and as pointed out in Section 1.4, it is supposed that the robot moves on narrow spaces. This physical limitation can be translated to error coordinates, since the current robot location is represented in terms of distance from the origin, which is the location of the reference robot. If the errors are maintained bounded and close to zero, then the behaviour of the real robot is close to

[3] Notice that it is also possible to consider the estimated slip, in spite of the nominal one. In such case, the estimated slip will be fixed during the prediction horizon. However, it is supposed that, in that case, the uncertainty set could be different from the one considered here.

the ideal one, and, therefore, far from the obstacles. In this sense, bounds on the states (e_x, e_y, e_θ) are established as[4]

$$\begin{aligned} e_x^m \leq e_x(k) \leq e_x^M, \\ e_y^m \leq e_y(k) \leq e_y^M, \\ e_\theta^m \leq e_\theta(k) \leq e_\theta^M. \end{aligned} \tag{2.31}$$

On the other hand, track motors have physical bounds in terms of maximum (or minimum) velocity. For that reason, as commented in Remark 4, velocity constraints must be translated to virtual control signals constraints[5] (constraints mapping). The constraints in terms of velocities are expressed as

$$\begin{aligned} d_r^m \leq C_r u(k) \leq d_r^M, \\ d_l^m \leq C_l u(k) \leq d_l^M, \end{aligned} \tag{2.32}$$

where $C_r = [-1 \quad -b/2]$, $C_l = [-1 \quad b/2]$, and

$$\begin{aligned} d_r^M &= \left(v_r^M(1 - i_r^M)\right) - v_r^{rf,M}, \\ d_r^m &= \left(v_r^m(1 - i_r^m)\right) - v_r^{rf,m}, \\ d_l^M &= \left(v_l^M(1 - i_l^M)\right) - v_l^{rf,M}, \\ d_l^m &= \left(v_l^m(1 - i_l^m)\right) - v_l^{rf,m}. \end{aligned} \tag{2.33}$$

Notice that state and input constraints can be considered in terms of

$$e(k) \in E, \quad u(k) \in U, \tag{2.34}$$

where $E \subseteq \mathbb{R}^3$ and $U \subseteq \mathbb{R}^2$ are polytopes that contain the origin.

2.6 Estimating Slip

This section describes some typical ways to estimate slip (2.1) and how two of them have been adapted and implemented for the physical experiments in this book.

First, some techniques to estimate the slip are reviewed. The most important differences come from how the forward velocity of the vehicle and the angular velocities of the tracks are determined. For instance, in [9], angular velocity is measured using two kinds of sensors, a proximity sensor and

[4] Recall from Section 1.4 that such bounds will be imposed according to each particular scenario.

[5] Notice that, for notational convenience, the term *virtual control signals* is also referred to u_1 and u_2. Strictly speaking, this term should be only applied to the control signals obtained through the dynamic feedback linearization procedure (2.20)-(2.21).

a dynamo-tachometer sensor. The linear velocity of the vehicle is obtained using three types of sensors, a cable-extension transducer, an encoder wheel, and a Doppler radar. In [2] and [50], slip is calculated as the difference between the velocity obtained using visual data and the kinematic model of the vehicle. A method for lateral slip estimation based on visually observing the trace produced by the wheels is presented in [112]. In the work [116], slip is estimated using a Kalman filter combining information from an Inertial Navigation System (INS) and laser scanner sensors. An Extended Kalman Filter (EKF) combining IMU, GPS and wheel encoder measurements is proposed in [130]. In [80], the authors use the difference between position measured by GPS and foreseen position computed from kinematic model to estimate slip. For a review of more slip estimation strategies see [60].

In this book, a first solution was to measure the angular velocity of the tracks using encoders and the robot forward velocity using a Doppler radar. The main advantage of this approach is that it is a straightforward approach to estimate the slip. For instance, solutions based on consumer-grade GPS provide low frequency updates making GPS too slow for slip estimation when fast applications are being employed [60], as it is the case of this book. The main limitation of this first strategy comes from the Doppler radar. Particularly, the Doppler radar suffers multi-path interference problems. Furthermore, it does not work properly at low velocities.

On the other hand, a similar approach to that proposed in [2] and [50] has also been implemented. In this case, the forward velocity of the robot comes from analyzing visual data using the technique known as *visual odometry* (see Chapter 3 for details). The angular velocity, as in the former case, comes from encoders. The main advantage of this approach is that it is an inexpensive method, since a consumer-grade camera replaces the Doppler radar. Additionally, it works properly when the robot moves at low velocities. The main limitation of this technique is that the performance of the vision-based algorithm is degraded in featureless environments and false matches can lead to erroneous slip estimations. These two approaches to estimate slip are analyzed in the following section through experimental tests.

2.7 Results

In this section, the extended kinematic model (2.5) is validated through several physical experiments. For that purpose, the TMR *Fitorobot* was used. Furthermore, an experimental method to estimate the additive uncertainty set is presented.

2.7.1 Testing the Sensor Performance

As commented above, two approaches to estimate slip have been implemented. Recall that the way in which slip is considered in this book follows as the relation defined in (2.1). For that reason, it is necessary to measure the forward linear velocity (v) and the theoretical velocity of each track ($\rho\omega_r$, $\rho\omega_l$). The theoretical velocity is measured using two incremental encoders attached to the track drive sprockets (DRS61, Sick AG, Waldkirch, Germany). The forward velocity has been measured by using two different alternatives, a Doppler radar (Compact II, LH Agro, Springfield, USA) and a vision-based approach called *visual odometry* (see Chapter 3 for details). Before validating the EKM, the tests carried out to check the performance of these two approaches are discussed.

Figure 2.4 shows the result of an experiment moving the robot at velocities in the range [0, 1] [m/s]. In this case, the theoretical velocities obtained from encoders are also plotted (labeled as "Right enc." and "Left enc."). It is observed that the Doppler radar does not work properly at velocities lower than 0.4 [m/s]. This erroneous behaviour is due to the resolution of the sensor. The vision-based approach (referred to as "Visual Odometry") works properly for low velocities. However, it produces a unsatisfactory result for

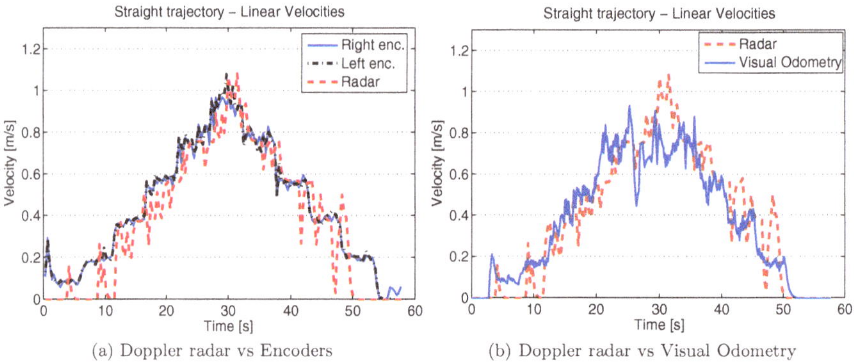

(a) Doppler radar vs Encoders (b) Doppler radar vs Visual Odometry

Fig. 2.4 Linear velocities during an experiment on pavement terrain

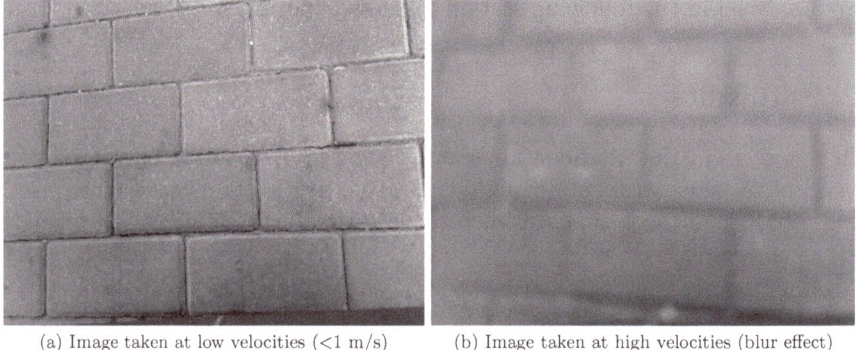

(a) Image taken at low velocities (<1 m/s) (b) Image taken at high velocities (blur effect)

Fig. 2.5 Images used for the vision-based approach

velocities close to 1 [m/s]. It was observed that when the robot moves at these velocities, a blur effect occurs in the images, see Figure 2.5. This issue leads to unsuccessful results and it will be deeply analyzed in Chapter 3, Section 3.3.

Once it was advised that Doppler radar works properly at middle-high velocities (> 0.4 [m/s]), it was also tested for a circular trajectory. Notice the different velocities of the encoders in Figure 2.6a. Again, the Doppler radar works correctly after a transient, that is, after 3 [s]. The vision-based approach was tested in a low-velocity experiment.

As verified in Figure 2.6b, it works properly for the range [0, 0.4] [m/s].

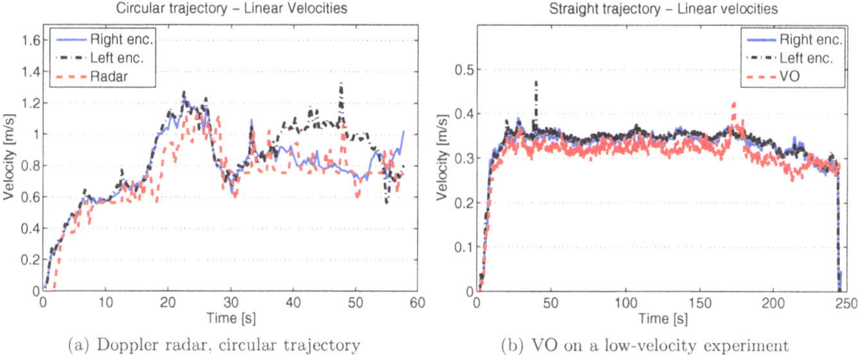

(a) Doppler radar, circular trajectory

(b) VO on a low-velocity experiment

Fig. 2.6 Physical experiments testing the performance of the Doppler radar and the vision-based technique to estimate the linear robot velocity

After many physical experiments, the conclusion was that the Doppler radar cannot be employed at low velocities (< 0.4 [m/s]), since it does not work properly or produces shaky readings. Notice that similar problems to these detected here are also reported in [9], [60]. On the other hand, the vision-based approach works satisfactorily for velocities lower than 1 [m/s]. For higher velocities, blur effect affects to the images and the forward velocity estimation becomes erroneous.

2.7.2 Model Validation

In this subsection, the EKM (2.5) is validated and compared with the CKM (2.3). Notice that model (2.5) has been discretized using a sampling period of $T_s = 0.25$ [s]. In order to compare both models, the mobile robot *Fitorobot* was teleoperated over different slip-behaviour flat terrains for a fixed distance of 20 [m]. Notice that it can be considered a reasonable distance to obtain reliable conclusions from the physical experiments. The velocity was ranged between [0, 1.2] [m/s], the track radius is $\rho = 0.15$ [m] and the distance between the track centres is $b = 0.5$ [m]. It is important to remark that the robot was moving on open-loop experiments (manually driven), what means that the error between the reference and the trajectory obtained from the kinematic models will diverge. Even the ground-truth can slightly differ with the marked reference trajectory. Furthermore, since the robot mainly moved to velocities greater than 0.6 [m/s], the Doppler radar was selected to measure the linear forward velocity.

The data from a DGPS (R100, Hemisphere, Calgary, Canada) was recorded, which is used as ground-truth. The DGPS has an accuracy of 0.20 [m]. Notice that in order to validate the trajectories obtained with the EKM and the CKM, the global position (Latitude/Longitude) has been converted to Universal Transverse Mercator (UTM) grid system [51], and hence to relative position from the starting point. On the other hand, the sensor data were filtered by a low-pass filter, with a smoothing factor (filter constant) of 0.78 [s] for the encoders and the vision-based measurements, and of 0.70 [s] for the Doppler radar readings.

First, the mobile robot was moved on a sandy terrain, see Figure 2.7a. It constitutes a deformable terrain in which tracks suffer slip and sinkage phenomena, see Figure 2.7b. Figure 2.7c shows a comparison of the trajectories obtained using the kinematic models. In this case, the trajectory obtained using the EKM closely follows to the ground-truth obtained from DGPS. The Euclidean distance between the travelled trajectory using the EKM and the ground-truth is 0.73 [m], and using the CKM is 2.78 [m]. That means an error of 3.68 [%] with respect to the total travelled distance for the EKM and of 14.03 [%] for the CKM. As expected, since the CKM does not take into account the slip phenomenon, the travelled distance is larger than that reached by the mobile robot. These data show that slip has a great influence in the robot motion. Finally, Figure 2.7d displays the slips of each track. In this experiment, the median slip value was close to 12 [%].

It is important to remark that although a high value of slip is expected according to similar experiments appeared in the literature, see for instance [2], different reasons justify the obtained result: (i) experiments similar to those carried out here are related to wheeled robots. In this case, a tracked robot was employed, and these vehicles are typically built to reduce slip and sinkage effects, what means that slip will be smaller than in wheeled vehicles in the same conditions, see for instance [54], [114], [133]; (ii) The robot was moved on a completely flat terrain, in the literature the experiments are usually carried out in loose terrains with significant slopes [2], [62].

It is also important to remark that before 1.5 [s], the Doppler radar measurements are zero due to the resolution of the sensor. For that reason, slip at the beginning of the test is 100 [%].

Secondly, the mobile robot was moved on gravel soil, see Figure 2.8a. This soil is composed of a thin stony/sandy top layer with a rigid compacted sand layer below, see Figure 2.8b. As checked in Figure 2.8c, the trajectory

(a) Mobile robot (b) Detail of the soil

(c) Travelled distances (d) Slip (Doppler radar and encoders)

Fig. 2.7 Model validation on sand soil

obtained using the kinematic models does not follow the ground-truth due to the open-loop nature of the experiments. Nevertheless, the EKM follows better the ground-truth than the CKM. The Euclidean distance between the travelled trajectory using the EKM and the ground-truth is 2.18 [m], and using the CKM is 5.46 [m]. That means an error of 10.58 [%] with respect to the total travelled distance for the EKM and of 26.52 [%] for the CKM. Again, the travelled distance obtained using CKM largely deviates from the ground-truth due to slip effect. Figure 2.8d shows the slips of each track. In this case, the median slip value is 4.5 [%]. As expected, the slip is lower than in sandy terrain, the reason is that this gravel soil constitutes a rigid terrain where there is a small sinkage effect.

Afterwards, the mobile robot was moved on grass, see Figure 2.9a. This terrain has a grassy top layer with a soft sandy layer below, see Figure 2.9b. The trajectories are compared in Figure 2.9c. The trajectory obtained using the EKM closely follows the ground-truth. The Euclidean distance between the travelled trajectory using the EKM and the ground-truth is 1.40 [m], and using the CKM is 3.64 [m]. That means an error of 7.24 [%] with respect to

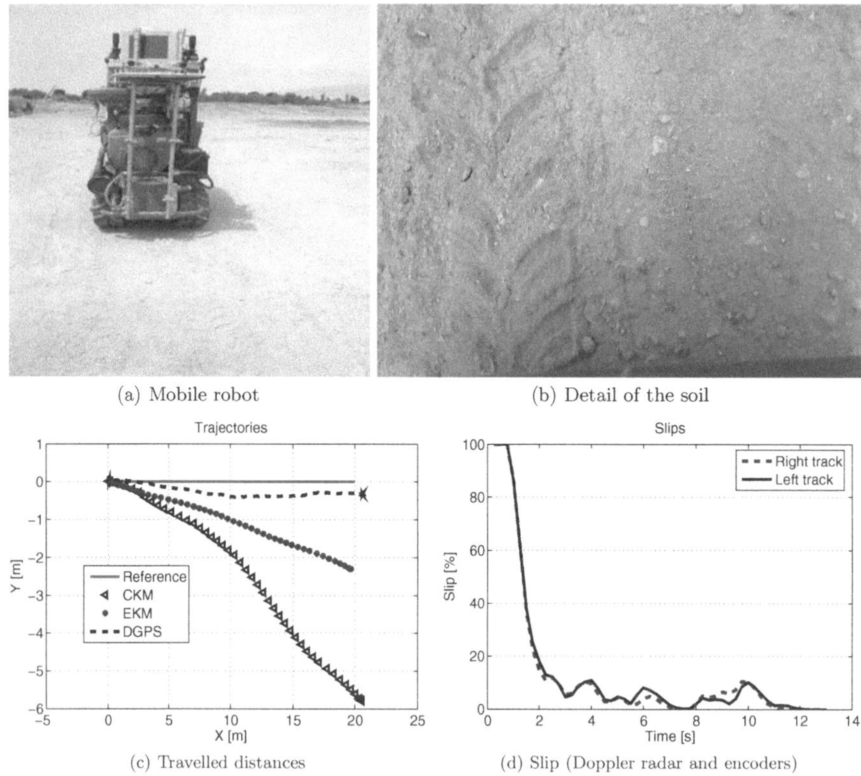

Fig. 2.8 Model validation on gravel soil

the total travelled distance for the EKM and of 18.84 [%] for the CKM. The slips of each track are depicted in Figure 2.9d. In this case, the median slip value is 8 [%]. As expected, the slip is lower than in sandy terrain. However, it is higher than that obtained in gravel soil. The reason is that the terrain below the grassy layer is softer than that of the gravel soil. This induced a greater sinkage effect and, hence, a greater slip.

Finally, the mobile robot was moved on a pavement ground, see Figure 2.10a. In this case, the robot suffered a quite low slip and an insignificant sinkage, since it is a rigid non-deformable surface, see Figure 2.10b. Figure 2.10c shows a comparison of the trajectories obtained using the kinematic models. The Euclidean distance between the travelled trajectory using the EKM and the ground-truth is 1.17 [m], and using the CKM is 2.93 [m]. That means an error of 5.70 [%] with respect to the total travelled distance for the EKM and of 14.29 [%] for the CKM. Although the longitudinal travelled distance obtained using the EKM is smaller than the ground-truth, the Euclidean distance between the ground-truth and the trajectory obtained using the EKM is smaller than in the case of the CKM. Figure 2.10d displays the slips of

(a) Mobile robot (b) Detail of the soil

(c) Travelled distances (d) Slip (Doppler radar and encoders)

Fig. 2.9 Model validation on grass soil

each track. In this case, the median slip value is 1.15 [%]. As expected, the slip is very small due to the robot was moving on a non-deformable surface.

In conclusion, Table 2.1 summarizes the most important data from the physical experiments carried out to compare the EKM with the CKM. As previously remarked, notice that smaller deviation between the trajectory obtained using the EKM and the ground-truth in relation to the CKM. Especially remarkable, the Euclidean distance on sandy soil is almost four times smaller than in the CKM case.

2.7.3 Additive Uncertainty Identification

The set $W \subseteq \mathbb{R}^3$ defined in (2.28) represents the uncertainty affecting the state at each sampling instant. This uncertainty bounds the mismatch between the continuous-time non-linear trajectory tracking error model and the discrete-time linear model, the noise in the slip estimation, and the uncertainty in the robot localization (see Section 2.4).

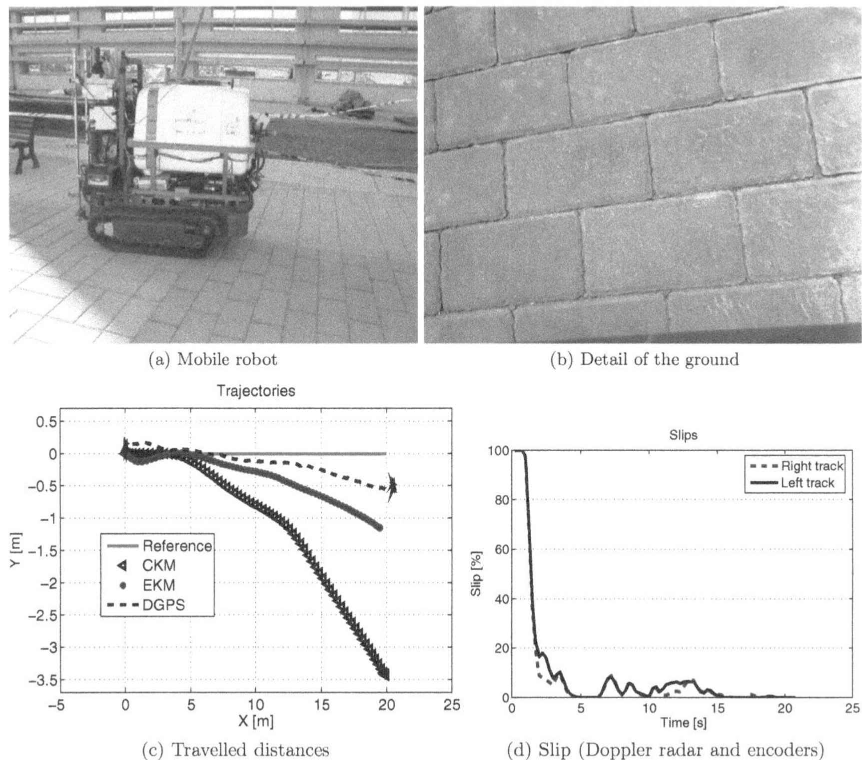

(a) Mobile robot (b) Detail of the ground

(c) Travelled distances (d) Slip (Doppler radar and encoders)

Fig. 2.10 Model validation on pavement ground

Table 2.1 Summary of model validation experiments

Feature / Surface	*Sand*	*Gravel*	*Grass*	*Pavement*
Median Slip [%]	12	4.5	8	1.15
Euclidean dist. EKM vs ground-truth [m]	0.73	2.18	1.40	1.17
Euclidean dist. CKM vs ground-truth [m]	2.78	5.46	3.64	2.93

This subsection focuses on the mismatch between the continuous-time non-linear trajectory tracking error model (2.10) and the discrete-time linear model (2.26). For that purpose, different input values for both models have randomly simulated. The set W has been obtained as the bounded mismatch between these models. Notice that $w \in W$ will be a vector of the form $[w_x \quad w_y \quad w_\theta]^T$, where each component is related to the mismatch in the longitudinal (e_x), lateral (e_y), and orientation (e_θ) error state, respectively. It is important to point out that such set, W, will be heuristically enlarged to take into account slip estimation noise and uncertainty in robot localization (see Chapter 5).

In this case, 1000 random points using the bounds defined in Table 2.2 have been simulated, and supposing that the distance between the track centres is $b = 0.5$ [m] and the sampling period is $T_s = 0.1$ [s]. As an example, Figure 2.11 shows the random values for the states and the reference velocities. Figure 2.12 presents the result of the simulation. From this experiment, it was checked that w_x is within [-0.028, 0.023] [m], w_y is within [-0.021, 0.022] [m] and w_θ is within [-0.97, 1.03] [deg].

Table 2.2 Bounds used to calculate the additive uncertainty term

	Max	Min	Units
Robot vel.			
v_r	2	-2	[m/s]
v_l	2	-2	[m/s]
States			
e_x	0.5	-0.5	[m]
e_y	0.5	-0.5	[m]
e_θ	20	-20	[deg]
Reference vel.			
v_r^{rf}	1.4	0.1	[m]
v_l^{rf}	1.4	0.1	[m]
Slip			
i_r	5	25	[%]
i_l	5	25	[%]

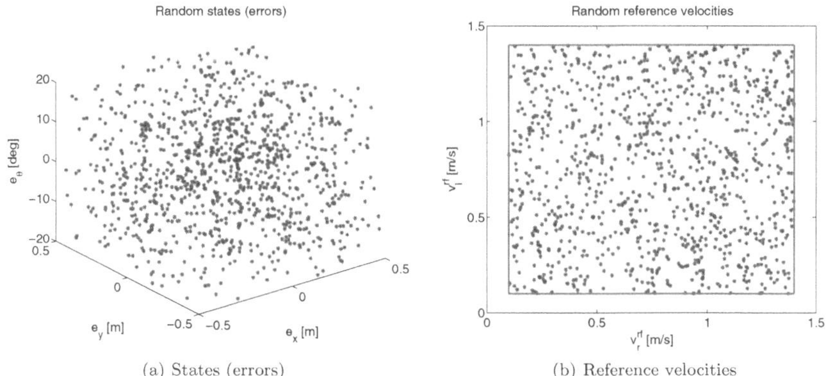

(a) States (errors) (b) Reference velocities

Fig. 2.11 Random values for the simulation. Simulation of 1000 points

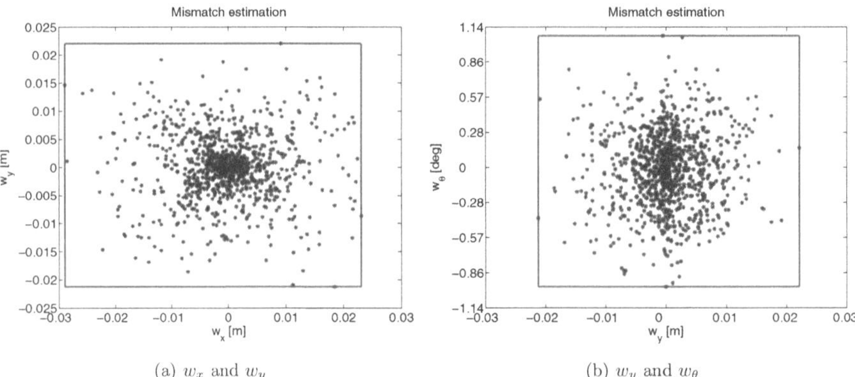

(a) w_x and w_y (b) w_y and w_θ

Fig. 2.12 Mismatches between the continuous-time non-linear EKM and the discrete-time linear EKM. Simulation of 1000 points

2.8 Conclusions

This chapter has presented several formulations to the trajectory tracking error model for a tracked mobile robot. The main contribution is that this model has been formulated using an extended kinematic model in which slip effect is taking into account. Such models will be used in Chapters 4 and 5 to synthesize several motion control strategies.

A discrete-time linear trajectory tracking error model including additive uncertainties has also been suggested. Such model is really advisable for mobile robotics applications where many uncertainties may occur. For instance, inaccurate robot location and noisy sensor measurements, among others. In this book, such uncertainty term deals with the mismatch between the continuous-time non-linear trajectory tracking error model and the discrete-time linear model, the noise in the slip estimation, and the uncertainty in the robot localization.

Another interesting conclusion from this chapter is the validation of two ways to measure the actual forward robot velocity and, hence, the slip. First, the forward velocity was measured using a Doppler radar. The main advantage of this approach is that it is an easy and immediate approach to estimate the slip. However, as checked in physical experiments, it does not work properly at velocities lower than 0.4 [m/s]. Secondly, the forward velocity of the robot was obtained analyzing visual data using the visual odometry technique. The main advantage is that it works properly when the robot moves at velocities lower than 1 [m/s]. The main limitation of this technique is the blur phenomenon.

The results carried out validating the EKM indicate that slip is especially noticed on sandy terrains where there is a certain degree of deformability. Particularly, the experiments carried out show a slip value around 12 [%]. Physical experiments evidenced that, for the case of mobile robots operating in off-road conditions, slip constitutes a unavoidable phenomenon, and an essential factor that affects the robot mobility. For that reason, appropriate localization techniques and control strategies that compensate or minimize its effects must be considered to achieve satisfactory and reliable results.

More precise estimations of the uncertainty set are a possible objectives of future works. Furthermore, new ways to estimate slip can be considered as further research. For instance, a proper approach can be to use one camera in order to estimate the actual velocity of each track.

Chapter 3
Localization of Tracked Robots in Planar Off-Road Conditions

This chapter presents the design of a visual-odometry-based localization technique. Typical implementation of visual odometry with one camera has been improved using another camera to estimate the robot orientation (visual compass approach). Comparisons with typical wheel-based odometry, through physical experiments, show the suitability of the proposed strategy for tracked mobile robots.

3.1 Introduction

Robot localization is defined as the process in which a mobile robot determines its current position and orientation relative to an inertial reference frame [124]. In the context of off-road mobile robots, localization techniques have to deal with the particular features of off-road conditions, such as a noisy environment (vibrations when the robot moves, disturbance sources, etc.), changing lighting conditions, high degrees of slip, and other inconveniences and disturbances.

One of the most popular solutions for the mobile robotics community is wheel-based odometry (or odometry) [14, 74]. This technique is considered as relative or local localization, that is, robot location is incrementally calculated from an initial point. Odometry employs simple geometric equations (mobile robot kinematics) with wheel encoders that provide angular velocities of the wheels. Then, the position and orientation is calculated by integrating these velocities. The main drawbacks of using wheel-based odometry are: (i) since encoder measurements are integrated, the noise is also integrated, and thus it causes a unbounded growth of the error along the time and the distance; (ii) it is based on the assumption that wheel revolutions can be converted into linear displacement relative to the terrain where, such as discussed in Chapter 2, and this assumption is limited on slip conditions (see equation (2.4)).

R. González, F. Rodríguez, and J.L. Guzmán, *Autonomous Tracked Robots in Planar Off-Road Conditions*, Studies in Systems, Decision and Control 6,
DOI: 10.1007/978-3-319-06038-5_3, © Springer International Publishing Switzerland 2014

An attractive alternative is the use of absolute or global techniques. These techniques determine the position of the robot with respect to a global reference frame, for instance, using beacons or landmarks [30, 118]. The most popular technique is GPS, which is based on satellite signals to determine the absolute position of an object on the Earth (longitude, latitude and altitude) [51]. The main drawbacks of absolute techniques are: (i) it requires a costly installation of the beacons/markers on the area where the robot operates, (ii) the mobile robot can only navigate over the area in which landmarks are located. Furthermore, the particular problems related to GPS are: (i) the satellite signal is lost in partially covered areas (nearby trees, buildings, etc.); (ii) it cannot be used in covered areas (greenhouses, mines, etc.) or in space exploration [51]; (iii) consumer-grade GPS provides poor accuracy (several meters). Although more expensive solutions such as Differential GPS (DGPS) or Real-Time Kinematics GPS (RTK-GPS) improve significantly that accuracy.

On the other hand, techniques that estimate robot location using visual information (images) are being successfully applied to off-road mobile robots, especially in space robotic exploration missions [23, 63, 92, 105]. One of the most popular approaches is *visual odometry*, which is defined as the incremental online estimation of robot motion from image sequences using an on-board camera [19, 91].

Finally, many efforts are being developed at the probabilistic estimation techniques field. In the mobile robotics context, probabilistic techniques are based on estimating the location of a robot combining measurement data from different sources, and prior knowledge about the system and measuring devices. The most representative approaches are based on the Kalman filter [93, 131] and the Particle filter [123]. Particularly, these techniques have been fruitfully applied to the Simultaneous Localization and Mapping problem[1] (SLAM) [29, 122, 123].

This chapter describes the work related to the application of a visual odometry approach (combined with visual compass) to off-road mobile robot localization. In particular, to tracked mobile robots operating in planar off-road environments. Comparisons of this strategy with typical localization techniques, through physical experiments using a real tracked robot, show the satisfactory behaviour of the proposed scheme.

This chapter is organized as follows. In Section 3.2, the localization strategy based on visual odometry is detailed. Physical experiments are discussed in Section 3.3. Finally, Section 3.4 is devoted to conclusions and future research.

[1] The SLAM problem deals with starting from an arbitrary initial point, a mobile robot should be able to autonomously explore the environment with its on-board sensors, gain knowledge about it, interpret the scene, build an appropriate map, and localize itself relative to this map [118].

3.2 Localization Using Visual Odometry

This section presents the work related to the application of a visual odometry approach to estimate the location of a mobile robot operating in off-road conditions.

Visual odometry is defined as the incremental online estimation of robot motion from an image sequence by an on-board camera [19, 91]. Roughly speaking, it is based on estimating the robot motion through the displacement of the objects found in two consecutive acquired images. The most important features of visual odometry in mobile robotics are:

- It is a straightforward technique. Visual odometry follows the same concepts of simplicity and practicality of wheel-based odometry. In fact, simple principles of odometry can be successfully applied to other robot configurations where wheel encoders are not available, such as humanoid robots [110] or unmanned aerial robots [102].
- A consumer-grade camera can replace a typical expensive sensor suite (encoders, IMU, GPS, etc.). It is really advisable for low-cost robots and/or with a limited payload.
- It avoids typical multi-path interference problems as in ultrasonic sensors and low-cost infrared sensors [19].
- For robots operating in indoor environments, visual odometry is significantly more practical and functional than other localization systems [3, 19].
- For robots in outdoor scenarios, slip effect is minimized, since visual information gives the actual robot velocity [2, 50].

The main limitations related to visual odometry are:

- Similar to wheel-based odometry, the robot location is integrated, and thus it will lead to error growth along time and distance.
- The visual odometry success depends highly on the light and imaging conditions (i.e., terrain appearance, shadows, camera parameters, etc.). Visual odometry can fail in dark or bad illuminated environments.
- It demands a middle/high computation cost, so appropriate computation resources must be employed.
- If the environment is completely featureless, the technique can fail. On uniform soils, such as checkerboard-type, visual odometry can also fail, since different places will produce similar images.
- If the mobile robot moves at high velocities (> 1 [m/s], in the particular case of this book) blur phenomenon can corrupt the images, which can lead to an unsuccessful result. This point can be minimized by selecting a camera with a proper exposure time.

Generally, there are two ways to estimate the location of a mobile robot using the visual odometry paradigm. The most popular method is called *optical flow* [85, 86, 91] (see Figure 3.1a). It is based on tracking distinctive features between successively acquired images [85]. In this case, an image is matched with the previous one by individually comparing each feature on them and finding candidate matching features based on Euclidean distance of their feature vectors. Afterwards, the velocity vector between these pairs of points is calculated and the displacement is obtained by using these vectors [85]. Optical flow is especially advisable for textured scenarios, such as urban and rough environments [23, 63, 115]. This approach has been tested using single [102], stereo [90], and omnidirectional cameras [115].

A slightly different approach is the *template matching* method [17, 44, 73] (see Figure 3.1b). It avoids the problem of finding and tracking features, and instead it looks at the change in the appearance of the world (images). For that purpose, it takes a template or patch from an image and tries to match it in the previous image. The main difference with optical flow is that now, no identification or tracking of features is involved, and there is no need to measure image velocities at different locations [119]. Appearance-based method has been successfully applied employing single [65, 104] and omnidirectional cameras [73].

Figure 3.1 shows two examples of the application of optical flow and template matching approaches. In the case of Figure 3.1a, the squares indicate the tracked features and the lines show the motion of these features from the previous image. In Figure 3.1b, the blue rectangle means the template in the previous image, the black rectangle is the result of the matching process with the previous image, and the green line indicates the pixel displacement between both images.

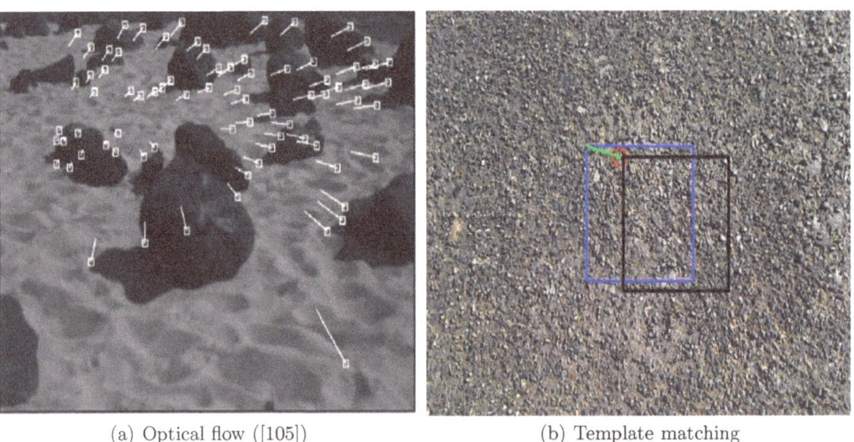

(a) Optical flow ([105]) (b) Template matching

Fig. 3.1 Example of visual odometry approaches: optical flow and template matching

Now, the most proper visual odometry implementation for off-road conditions is analyzed. The main difference between optical flow and template matching approaches is that when the scene is low textured, the number of detected and tracked features (single patterns) is low, what can lead to poor accuracy of motion estimate [63]. This fact means that optical flow can fail on featureless scenarios (such as sandy soils, urban floors, etc.) where images with few high gradients are grabbed. On the other hand, the template matching approach works properly in low texture scenarios, since a larger pattern (template) is employed, and, therefore, the probability of a successful matching is increased [73].

Previous discussion motivates why the template matching method has been selected in this book. However, it is important to remark that if the matching process fails (false matches), the robot motion estimate can become degraded. In order to minimize this shortcoming, especially undesirable estimating robot orientation, a second camera is added. This solution is inspired by two recent works, where a method called *visual compass* was proposed to estimate rotational information from omnidirectional cameras [73, 115]. Visual compass technique is based on the use of a camera mounted vertical to the ground on a mobile robot. Then, a pure rotation on its vertical axis result in a single column-wise shift of the appearance in the opposite direction. In this way, the rotation angle is retrieved by matching a template between the current image (after rotation) and the previous one (before rotation) [115].

In conclusion, the proposed localization strategy takes as input two image sequences. One image sequence comes from a single camera pointing at the ground under the robot ("groundcam"), and the second one comes from a camera looking at the environment ("pancam"). The former is employed to estimate the robot longitudinal displacement (Subsection 3.2.2), and the latter is employed to estimate the robot orientation (Subsection 3.2.3).

3.2.1 Template Matching

In this subsection, the mathematical formulation of template matching is briefly described.

The template matching method is defined as the process of locating the position of a sub-image inside a larger image. The sub-image is called the *template* and the larger image is called the *search area* [17, 44]. This process involves shifting the template over the search area and computing the similarity between the template and a window in the search area. This is achieved by calculating the integral of their product. When the template matches, the value of the integral is maximized.

There are several methods to address the template matching, see [16], [113], for a review. Here, the cross-correlation solution has been implemented[2]. It

[2] A trade-off was realized comparing different methods (e.g. square difference matching method, correlation matching method, etc.) using images from physical experiments. The best result (fewer false matches) was obtained using the cross-correlation approach.

is based on calculating an array of dimensionless coefficients for every image position (s, v) as [17, 113]

$$R(s,v) = \sum_{i=0}^{h-1} \sum_{j=0}^{\varpi-1} \big(T(i,j) - \bar{T}(i,j)\big)\big(I(i+s,j+v) - \bar{I}(i+s,j+v)\big), \quad (3.1)$$

where $h \in \mathbb{R}^+$, $\varpi \in \mathbb{R}^+$ are the height and the width of the template, respectively, $T(i,j)$ and $I(i,j)$ are the pixel values at location (i,j) of the template and the current search area, respectively, and $\bar{T}(i,j)$ and $\bar{I}(i,j)$ are the mean values of the template and current search area, respectively. These mean values are calculated as

$$\bar{T}(i,j) = \frac{1}{(\varpi h)} \sum_{a=0}^{h-1} \sum_{c=0}^{\varpi-1} T(a,c), \quad (3.2)$$

and

$$\bar{I}(i+s,j+v) = \frac{1}{(\varpi h)} \sum_{a=0}^{h-1} \sum_{c=0}^{\varpi-1} I(a+s,c+v). \quad (3.3)$$

Now, in order to avoid changes in the brightness between the template and the current image, every correlation coefficient is normalized [16]. For that purpose, it is divided by the standard deviation

$$N(s,v) = \sqrt{\sum_{i=0}^{h-1} \sum_{j=0}^{\varpi-1} \mathfrak{T}(i,j)^2 \cdot \mathfrak{I}(i+s,j+v)^2}, \quad (3.4)$$

where $\mathfrak{T}(i,j) = T(i,j) - \bar{T}(i,j)$ and $\mathfrak{I}(i+s,j+v) = I(i+s,j+v) - \bar{I}(i+s,j+v)$.

Finally, the normalized cross-correlation becomes

$$\tilde{R}(s,v) = \frac{R(s,v)}{N(s,v)}. \quad (3.5)$$

Notice that the value of $\tilde{R}$ changes between -1 and $+1$, and the closer $\tilde{R}$ is to $+1$, the more similar the template and the current image will be. For that purpose, the best match is defined as

$$\tilde{R}^M = max\big(\tilde{R}(s,v)\big), \quad (3.6)$$

where $\tilde{R}^M$ is the maximum value of the array $\tilde{R}$ and (s^M, v^M) is the position of that point.

3.2.2 Estimating Robot Displacement

This subsection focuses on the estimation of the robot longitudinal displacement using the images taken by the camera pointing at the ground.

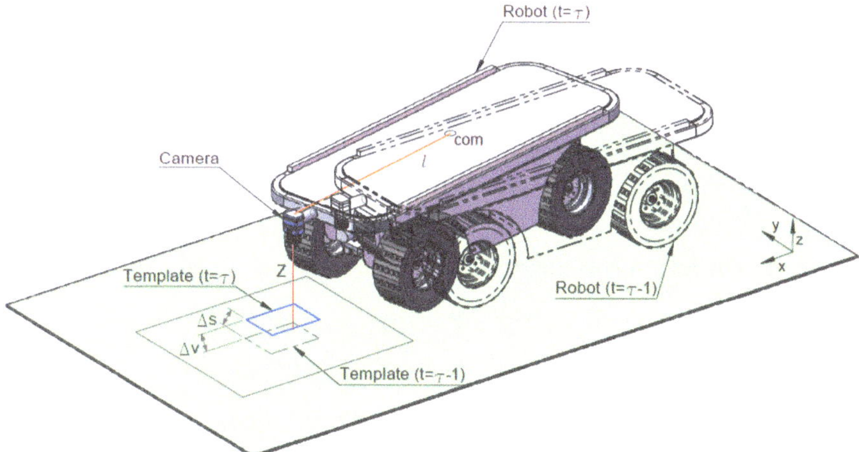

Fig. 3.2 Visual Odometry based on Template Matching using a camera pointing at the terrain under the robot

As shown in Figure 3.2, at sampling instant $t = \tau - 1$, the robot takes a picture of the ground under it. At the following sampling instant $t = \tau$, the template matching approach is employed to find a defined template from the previous image in the current image. Finally, the pixel displacement (Δs, Δv) is calculated as

$$\begin{aligned} \Delta s &= T(\tilde{i}) - i^M, \\ \Delta v &= T(\tilde{j}) - j^M, \end{aligned} \tag{3.7}$$

where $\Delta s \in \mathbb{R}$, $\Delta v \in \mathbb{R}$ are the longitudinal and lateral pixel displacements from the image sequence taken by the camera pointing at the ground, $T(\tilde{i}, \tilde{j})$ is the top left corner of the template (rectangle region centered at previous image), and (i^M, j^M) is the point of maximum correlation. Notice that, for notational convenience, the time dependence on previous variables has been omitted.

Afterwards, camera units must be translated to physical world units using the camera calibration parameters,

$$\begin{aligned} \Delta x &= \Delta s \frac{Z}{f_x^g}, \\ \Delta y &= \Delta v \frac{Z}{f_y^g}, \end{aligned} \tag{3.8}$$

where $\Delta x \in \mathbb{R}$, $\Delta y \in \mathbb{R}$ are the camera longitudinal and lateral displacements in physical world units, respectively, $Z \in \mathbb{R}^+$ is the height of the camera above ground (see Remark 3), and $f_x^g \in \mathbb{R}$, $f_y^g \in \mathbb{R}$ are the focal lengths of the camera pointing at the ground.

Assumption 3. *It is assumed that the robot is moving on a predominant flat ground and sinkage does not change, what means that the distance between the camera and the ground is almost constant. Notice that, on a rougher surface, an IMU sensor or a laser sensor should be used to estimate this distance [104]. Recently, a novel approach consists in the use of telecentric cameras [101]. Those cameras are electronically modified in such a way that the lens keep the same field of view, regardless of the distance between the camera and the ground.*

Finally, the location of the robot along the time is given by (see Remark 5)

$$\begin{aligned} x^{vo}(k) &= x^{vo}(k-1) + \Delta x(k)\cos\left(\theta^{vo}(k)\right), \\ y^{vo}(k) &= y^{vo}(k-1) + \Delta x(k)\sin\left(\theta^{vo}(k)\right), \end{aligned} \tag{3.9}$$

where $[x^{vo} \quad y^{vo}]^T \in \mathbb{R}^2$ is the robot position. The estimation of the robot orientation ($\theta^{vo} \in \mathbb{R}$) is addressed in the following subsection.

Remark 5. *Notice that the robot orientation can be calculated using the information from the camera pointing at the ground. In this case, it is obtained as [104]*

$$\Delta\hat{\theta} = \arctan(\Delta y, l), \tag{3.10}$$

where $\Delta\hat{\theta} \in \mathbb{R}$ is the increment in robot orientation, and $l \in \mathbb{R}^+$ is the distance between the camera and the robot centre (see Figure 3.2). Then, the orientation at each sampling instant is given by

$$\hat{\theta}^{vo}(k) = \hat{\theta}^{vo}(k-1) + \Delta\hat{\theta}(k). \tag{3.11}$$

However, the resulting orientation is extremely sensitive to systematic errors, such as inaccurate distance between the camera and the ground plane, inaccurate distance between the camera and the centre of the robot, and false matches. These drawbacks produce that orientation becomes less and less accurate at each step [115]. In order to minimize such effects, the visual compass technique is employed in this book.

3.2.3 Estimating Robot Orientation: Visual Compass

The application of the visual compass technique to calculate the robot orientation is explained in this subsection. The visual compass approach was recently presented as a new way to estimate the robot orientation using vision.

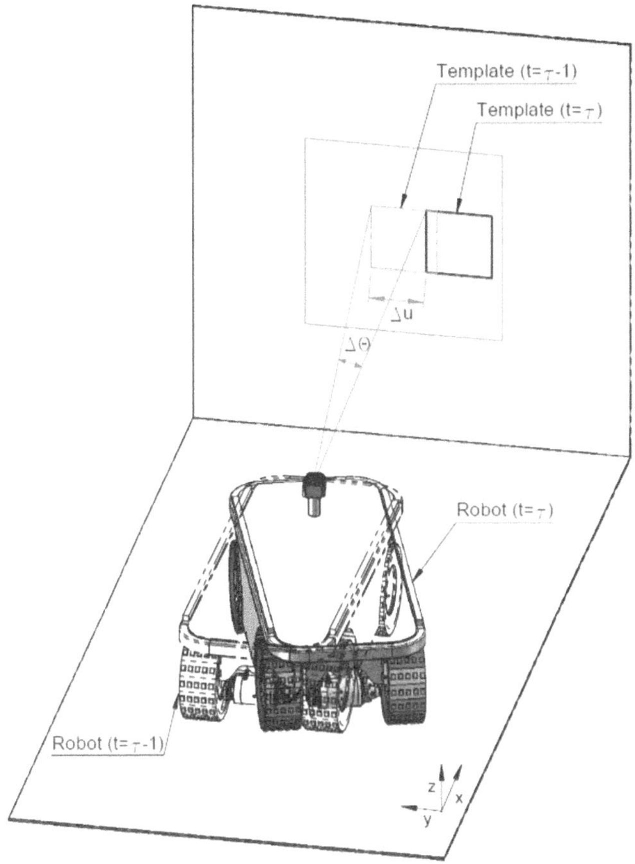

Fig. 3.3 Visual Compass approach using a camera looking at the environment

It was firstly presented in [73], and it has been mainly applied to omnidirectional camera systems [115, 120].

The visual compass technique is also based on the template matching procedure to estimate the pixel displacement between two consecutive images. The difference is that now, a camera looking at the environment ("panoramic view") is employed. In this way, a change in the robot orientation means a unidirectional pixel displacement between two consecutive images (see Figure 3.3). The procedure consists of firstly obtaining the maximum correlation point between both images using (3.6), and secondly, calculating the pixel displacement (only in one direction) between the top left corner of the template and the maximum correlation point, that is,

$$\Delta u = T(\tilde{i}) - i^M, \tag{3.12}$$

where $\Delta u \in \mathbb{R}$ is the pixel displacement from the image sequence taken by the camera looking at the environment. Finally, the rotation of the robot, $\Delta\theta \in \mathbb{R}$, supposing that the camera is mounted on the centre of the robot, is given by

$$\Delta\theta = \arctan(\Delta u, f_x^e), \tag{3.13}$$

$f_x^e \in \mathbb{R}$ being the focal length of the camera looking at the environment. Then, the orientation along time is given by

$$\theta^{vo}(k) = \theta^{vo}(k-1) + \Delta\theta(k). \tag{3.14}$$

Remark 6. *Notice that the book is based on using a TMR that can turn on the spot or with almost zero translations. This leads to consider that there is no translational motion at turns or it is close to zero, what implies that parallax effects are negligible [115].*

3.2.4 Localization Approach Combining Visual Odometry with Visual Compass

Summing up, the localization scheme presented here based on visual odometry and visual compass operates as follows:

1. Acquire a pair of consecutive frames from each camera.
2. Select the template from images taken at time $t = \tau - 1$.
3. Match the template with the current image ($t = \tau$) by using (3.1). Normalize the result by using (3.5).
4. Estimate the pixel displacement between the template and the maximum correlation point with (3.7).
5. Translate from camera plane to world plane using the camera calibration parameters by means of (3.8).
6. Compute the rotation angle using the visual compass method using (3.13).
7. Estimate the robot location using translation information given by the camera pointing at the ground with (3.9), and the rotation angle given by the camera looking at the environment with (3.14).
8. Repeat from step 1.

3.2.5 Computational Aspects of Template Matching

This subsection discusses some experiments carried out to select the most appropriate template/search area size for a satisfactory performance of the correlation algorithm and the employed computation time.

The main drawback of template matching approach is its computation cost, since the template has to be slid over the whole search area. In the general case, the detection of a single template $T_{m\times m}$ within a image $I_{n\times n}$

by means of a matching process is $\mathcal{O} = m^2(n - m + 1)^2$ [17]. For that reason, two important issues to be investigated are the template size and the search area size.

Notice that here, the possibility of speeding up the matching process from an algorithmic point of view is not being considered. This subsection only deals with determining the template/search area size to reach a trade-off between performance and computation time.

Speeding up the Matching Process by Reducing the Image Size

Firstly, the proper template size is studied, and later on, a way to reduce the search area is analyzed. In this way, the template is obtained as a reduced squared window of the image taken at sampling instant $\tau - 1$. The template origin has been established as the image centre, and the top left corner of the template is located at [104]

$$\begin{aligned} T^q(s) &= \frac{W^q}{2} - \frac{T^q_{size}}{2}, \\ T^q(v) &= \frac{H^q}{2} - \frac{T^q_{size}}{2}, \end{aligned} \tag{3.15}$$

where $T^q(s, v)$ is the top left corner of the template, q refers to the images taken by the camera pointing at the ground ($q = g$) and to the camera looking at the environment ($q = e$), $W^q \in \mathbb{R}^+$, $H^q \in \mathbb{R}^+$ are the width and height of the original image, respectively, and

$$T^q_{size} = \frac{1}{\rho^q} H^q, \tag{3.16}$$

is the template size being $\rho^q \geq 1$ a reduction factor.

Notice that the larger the template, the smaller the probability that it is matched in the search area. That means that if a too large template is selected, it cannot be possible to find it in the following image. On the contrary, the smaller the template, the higher the probability to fail into false matches. That is, if a too small template is selected, several areas of the following image can match with that template.

As previously commented, the second way to speed up the correlation matching process consists in to use a reduced window of the original image instead of the whole image. Such reduced search area is given by

$$\begin{aligned} Win^q_w &= \frac{1}{\lambda^q} W^q, \\ Win^q_h &= \frac{1}{\lambda^q} H^q, \end{aligned} \tag{3.17}$$

where Win^q is the size of the new reduced image, and $\lambda^q \geq 1$ is a reduction factor. Then, the reduced image will start at the point $Win^q(s, v)$ and it will have a size of $Win^q_w \times Win^q_h$. The top left corner of the new image is

$$Win^q(s) = Win^q_w - \frac{Win^q_w}{\lambda^q},$$
$$Win^q(v) = Win^q_h - \frac{Win^q_h}{\lambda^q}. \tag{3.18}$$

In this way, the computation time is decreased, since correlation process is carried out over a smaller image, such as shown in the following subsection.

Effect of Template and Image Sizes on the Computation Time

Before carrying out physical experiments, the effect of the template and image sizes on the computation time has been analyzed. Image sequences taken during physical experiments are also employed here (see Section 3.3). Notice that experiments have been carried out on a computer Intel Core 2 Duo 2.5 GHz with 3.5 GB RAM.

Figure 3.4 shows the resulting computation time of varying the template and image sizes ("Mean" is the mean computation time of the sequence of images and "Std" denotes the standard deviation). Here, the template and image sizes of the image sequence employed by the visual compass method are fixed to $\rho^e = 4$ and $\lambda^e = 1.7$ for (3.16) and (3.17), respectively. As observed, larger template size (smaller ρ^g) implies that the computation time is lower. When the matching process is applied over a smaller search area (larger λ^g) the computation time also decreases. The computation time when images do not have any reduction (black triangles) is also displayed. From this analysis, the following reduction factors $\rho^g = 3$ and $\lambda^g = 1.2$ have been selected, since they constitute a compromise between suitable computation time ($< 0.2[s]$)

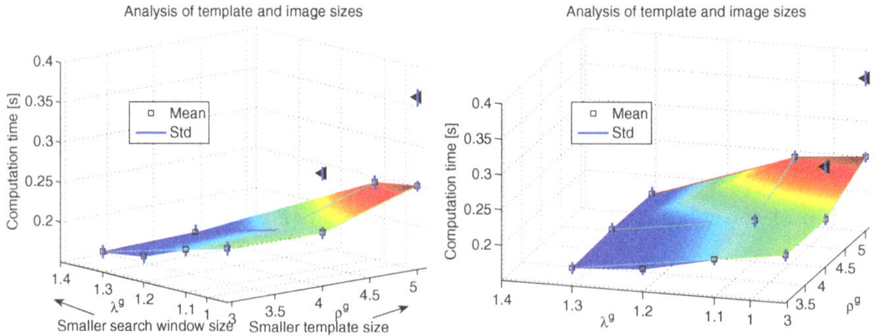

Fig. 3.4 Analysis of template and image sizes on computation time (images from camera pointing at the ground). The size of the images from the pancam is fixed

and success in the matching process. It is important to remark that although smaller reduction factors can be considered, these reduced search areas lead to a unfeasible matching process. That means that for the experiments carried out in this book if smaller search areas were considered, the number of false matches increases to unsuitable values, and robot location cannot be reliably estimated.

Notice that, such as remarked in [104], there is another important parameter to be considered in the selection of the template size, that is, the robot velocity. It is experienced that a smaller template size permits a high robot velocity, and a large template size limits the robot velocity. Regarding this issue, the images used for the experiment displayed in Figure 3.4 were collected for a robot velocity ranged between 0.4 and 0.5 [m/s] (see Section 3.3). Nevertheless, the selected template and image sizes still work properly for small variations of those velocities.

3.3 Results

As in previous chapter, physical experiments were carried out using the mobile robot *Fitorobot*. Additionally, successful results using the visual odometry approach were also obtained with the *CRAB rover* available at the ASL, ETH Zürich (Switzerland), see [39].

3.3.1 Testing the Sensor Performance

From physical experiments, it was noticed that when the robot moves at velocities greater than 1 [m/s] on flat terrains, blur phenomenon corrupts the images taken by the camera pointing at the ground, see Figure 3.5a. Additionally, it was observed that when the robot moves at velocities greater than 1 [m/s] on partially bumpy terrains, a similar effect corrupts again the images, see Figure 3.5b. Blur phenomenon occurs when an image is captured while the camera is moving during the exposure time or shutter time [110]. This phenomenon constitutes a difficult issue to be removed and elaborate solutions have to be considered to minimize its influence. For instance, in [55], authors formulate a learning policy as a trade-off between the localization accuracy and robot velocity. In [110], authors propose to carry out a preprocessing step before detecting features in the image. In this book, a preprocessing of the images, such as an enhancing filter, is not appropriate due to this would mean to raise the computation time assigned to the vision algorithm. The solution based on bounding the robot velocity can be successful. However, it would entail a certain degree of conservativeness for the motion controllers.

In relation to the image shown in Figure 3.5b, it is also interesting to remark that blur phenomenon is stressed by the vibrations affecting to the mobile robot. It is a difficult issue to be removed, since the TMR employed for physical experiments has a limited suspension mechanism that produces unavoidable vibrations on the robot structure. In conclusion, as a first approach in this book, visual odometry was employed when the robot was moving at velocities less than 1 [m/s]. Nevertheless, in the physical experiments presented in this book, the robot rarely reaches velocities greater than 1 [m/s].

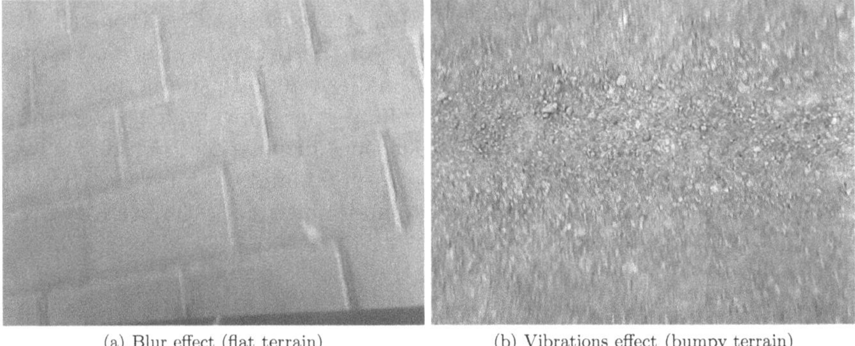

(a) Blur effect (flat terrain) (b) Vibrations effect (bumpy terrain)

Fig. 3.5 Images taken by the camera pointing at the ground (velocities > 1 [m/s])

Another important issue observed from outdoor physical experiments is the problem found in environments with changing lighting conditions, which can lead to shadows in the images taken by the camera pointing at the ground. After analyzing many experiments, it was concluded that when there are shadows in the images, the risk for false matches highly increases. For that reason, this phenomenon has been deeply studied and two approaches to minimize its effect have been proposed.

First, the position and the height of the camera pointing at the ground were carefully studied. In this case, the camera was mounted in front of the robot between both tracks at a height of 0.49 [m], see Figure 3.6. This distance was obtained as a trade-off between shadow-reduction and template matching performance, that is, higher distance leads to more features but shadows can appear. On the contrary, shorter distance corresponds to smaller field of view where the probability of shadows in the images is reduced. However, it can lead to featureless images.

Secondly, a threshold filter has been tuned. It compares the current pixel displacement with the previous ones, if the difference is greater than a threshold, then the current value is considered as an outlier. In this way, these peaks or outliers, due to false matches, are partially compensated. As shown in the following subsection, this filter works properly, and it requires a small computation time.

Fig. 3.6 Tracked mobile robot *Fitorobot* during physical experiments (notice the cameras Quickcam Sphere AF, Logitech)

3.3.2 Localization Strategies Validation

In this subsection, the physical experiments validating the localization approach presented in this chapter are discussed. To that end, the robot was manually driven on a gravel terrain (see Figure 3.7). Notice that this is a similar experiment site than that used in Chapter 2 (gravel soil). As evaluated in that chapter, slip is around 6 [%].

Several trajectories were tested. In this case, three experiments are selected. In the first one, the robot was driven along a squared trajectory of approximately 55 [m] long and 20 [m] width. The total travelled distance was close to 160 [m]. In the second experiment, the robot was driven along an S-shaped trajectory with three parallel paths to the x-axis of 80 [m] and two perpendicular paths of 20 [m]. The total travelled distance was close to 290 [m]. Finally, a circular trajectory is selected, the total travelled distance was close to 65 [m].

The orientation was collected using a magnetic compass (C100, KVH, Middletown, USA); for the vision-based localization technique, two consumer-grade cameras were employed (Quickcam Sphere AF, Logitech, Apples,

Switzerland). The magnetic compass has an accuracy of 0.1 [deg]. The rest of the sensors were the same than those used in Chapter 2, that is, a DGPS, two incremental track encoders, and a Doppler radar. The sampling period was $T_s = 0.2$ [s] and the robot velocity ranged within [0.4, 0.5] [m/s]. Recall from Chapter 2 that in this velocity range the Doppler radar as well as the visual odometry work properly.

In the experiments, the DGPS and the magnetic compass data were considered as ground-truth for position and orientation, respectively. Notice that, as in Chapter 2, the position obtained using the DGPS is translated to relative position. For that purpose, the global position (Latitude/Longitude) was converted to UTM grid system [51].

For comparison purposes, a wheel-based odometry localization approach was implemented.

Experiment 1. Squared Trajectory

In this experiment, the robot was manually driven on a sunlit illuminated gravel terrain following a squared trajectory. The lighting conditions did not produce any significant shadows during the experiment in this case.

Figure 3.7 shows two frames employed by the vision-based localization technique during this experiment. The pixel displacement is marked by the green line and the red circle, the template is labeled by the blue rectangle, and the black rectangle means the reduced area in which the matching process is carried out (see Subsection 3.2.5).

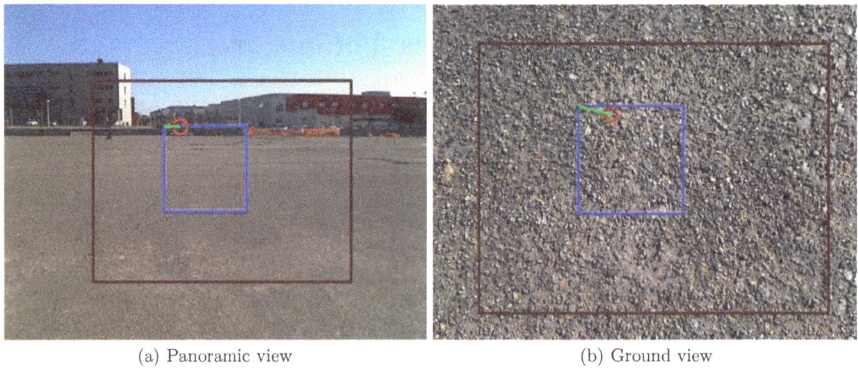

(a) Panoramic view (b) Ground view

Fig. 3.7 Result of template matching in the experiment site (gravel soil)

Figure 3.8a shows the resulting trajectories. It is observed that the visual odometry with visual compass trajectory (denoted as "VO+VC") closely follows the ground-truth (labeled as "DGPS"), while the wheel-based odometry estimate (denoted as "Odo") diverges largely from the ground-truth, particularly, odometry fails at turns. The trajectory obtained using the image sequence from the camera pointing at the ground (referred to as "VO") to estimate orientation (see Remark 5) is also plotted, and it has a similar result to that obtained using the approach combining information from both cameras (visual odometry with visual compass). Especially remarkable is that vision-based techniques work properly during straight-line motions and at turns. Figure 3.8b displays the orientations, where the ground-truth is denoted as "Compass". Here, it is checked that the orientations obtained through visual odometry fixes the ground-truth. The mean orientation errors are: 8.2 [deg] for visual odometry with visual compass, 14.1 [deg] for visual odometry, and 39.37 [deg] for wheel-based odometry. In this figure, it is possible to observe the unavoidable error growth phenomenon of odometry-based solutions, that is, the deviation between the ground-truth and the rest of techniques increases along the travelled distance.

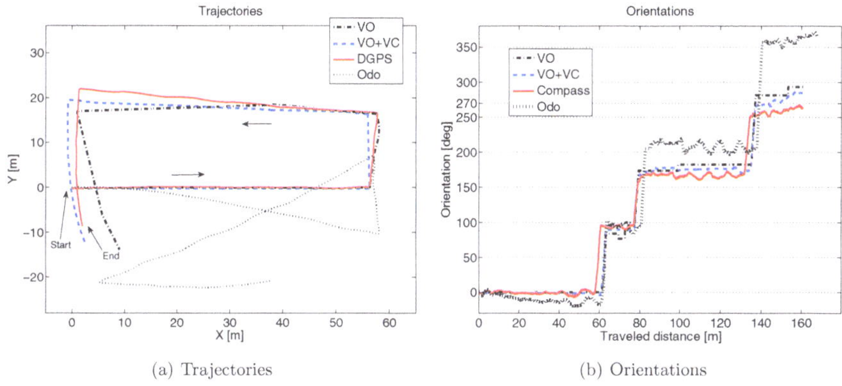

(a) Trajectories (b) Orientations

Fig. 3.8 Experiment 1. Squared trajectory

The error between each localization method and the ground-truth is also analysed quantitatively. In this case, the Euclidean distance between the initial and the latest position of the robot in the four parts of the trajectory is calculated, that is, the two parallel paths to the x-axis (Part 1 and Part 3) and the two perpendicular ones (Part 2 and Part 4). It is obtained that the visual odometry with visual compass approach achieves the smallest error. The other vision-based technique also achieves an admissible error. The relative mean errors with respect to the total travelled distance are: 1.45 [%] for visual odometry with visual compass, 2.33 [%] for visual odometry alone, and 16 [%] for wheel-based odometry.

In Figure 3.9, the longitudinal (Δs) and lateral (Δu) pixel displacement values related to the visual odometry with visual compass approach are shown. Notice that values close to zero mean small displacements (low velocity), and high values mean large displacements (high velocity). In this plot, it is observed that the points are aligned in two directions, being this effect due to the pixel displacements during straight motions, Δs component, and during turns, Δu component. It is checked that template matching is highly robust with few outliers (false matching or unsuccessful matching). It is important to point out three interesting conclusions from this plot. Firstly, since the robot always turns in the same sense (to the left side), lateral pixel displacement (Δu) is also aligned in one direction. Secondly, when the robot is turning it does not move forward, since, as observed, Δs is close to zero at turns. Finally, note that in the range $\Delta s = [-20, -40]$ [pixel], Δu is zero, what means the moment in which the robot is stopping before turning.

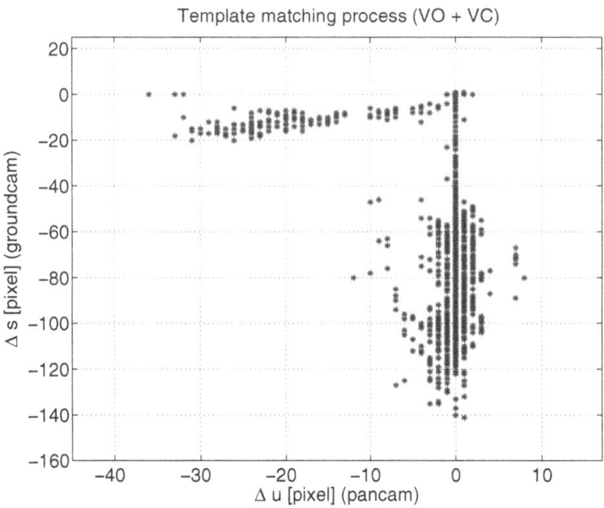

Fig. 3.9 Experiment 1. Lateral and longitudinal pixel displacements (VO+VC)

Experiment 2. S-shaped Trajectory

In this experiment, a longer trajectory, in which the robot changed several times of direction, was tested. Furthermore, in some parts of the experiment site, the lighting conditions produced shadows that affected to the performance of the vision-based localization strategies.

Figure 3.10a shows the resulting trajectories. It is observed that the visual odometry with visual compass trajectory does not follow accurately the ground-truth mainly for one reason. As checked during the first perpendicular path to the x-axis and the second parallel path, the trajectory is shorter than the ground-truth. This fact is due to false matches obtained from the camera pointing at the ground caused by shadows (see Figure 3.11b). This erroneous behaviour is worst in the case of visual odometry alone, since, now, longitudinal displacement and orientation are obtained from the camera pointing at the ground. The largest deviation is obtained during the first perpendicular path to the x-axis. Again, wheel-based odometry diverges largely from the ground-truth, particularly, odometry fails at turns. In Figure 3.10b, the orientations are plotted with respect to the travelled distance. Here, the erroneous behaviour of the visual odometry approach during the first parallel path to the x-axis is noticed. The visual odometry with visual compass approach estimates the orientation properly and follows the ground-truth. The mean orientation errors are: 4.8 [deg] for visual odometry with visual compass, 10.2 [deg] for odometry, and 148.2 [deg] for visual odometry. The mean orientation error for the case of visual odometry cannot be considered as a comparable value, since it has a large standard deviation.

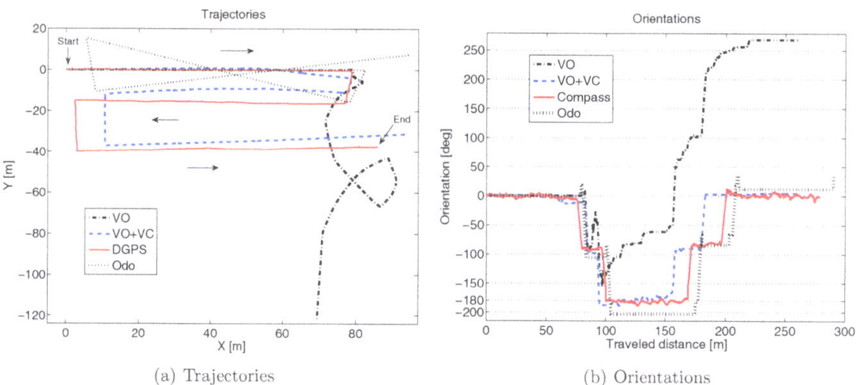

(a) Trajectories (b) Orientations

Fig. 3.10 Experiment 2. S-shaped trajectory

The error between each localization technique and the ground-truth is also obtained. In this case, the Euclidean distance between the initial and the latest position of the robot in five parts is calculated, that is, three parallel paths to the x-axis (Part 1, Part 3 and Part 5) and two perpendicular ones (Part 2 and Part 4). As expected, the visual odometry with visual compass approach obtains an admissible error except during the second and the third paths. The relative mean errors with respect to the total travelled distance are: 2.46 [%] for visual odometry with visual compass, 7.60 [%] for odometry, and 19.50 [%] for visual odometry.

In Figure 3.11a, the longitudinal (Δs) and lateral (Δu) pixel displacement values are shown. In contrast to the previous experiment, now, the pixels related to the lateral displacement (Δu) are aligned in two directions (right and left turns). Notice that the turns to the right side were carried out at higher linear velocity than the turns to the left side. In this case, there are some outliers, when the robot moved in straight line ($\Delta s < -40$ [pixel]), that can explain the small deviation obtained at the end of the first parallel path to the x-axis (see Figure 3.10). On the other hand, in Figure 3.11b, the longitudinal pixel displacements (Δs) with respect to the acquired images are displayed. As noticed, during the samples [1200, 1600], there is an erroneous behaviour (false matches). This behaviour explains why the trajectories obtained with the vision-based approaches are shorter than the ground-truth during the first perpendicular path to the x-axis. The false matches found in the interval [2300, 2800] explain why the trajectories are shorter than the ground-truth during the second parallel path to the x-axis.

A deeper understanding of the erroneous behaviour of the visual odometry approach is obtained analyzing the Figure 3.12. Recall that, for the case of visual odometry alone, the robot orientation comes from the lateral pixel displacement obtained from the camera pointing at the ground (see Remark 5). As checked in Figure 3.12a, many outliers appear in the Δv component, compare it with the visual compass pixel displacement (Δu) obtained from the camera looking at the environment in Figure 3.11a. From Figure 3.12b, notice that these outliers occur within the two intervals in which false matches appeared due to shadows (see Figure 3.11b).

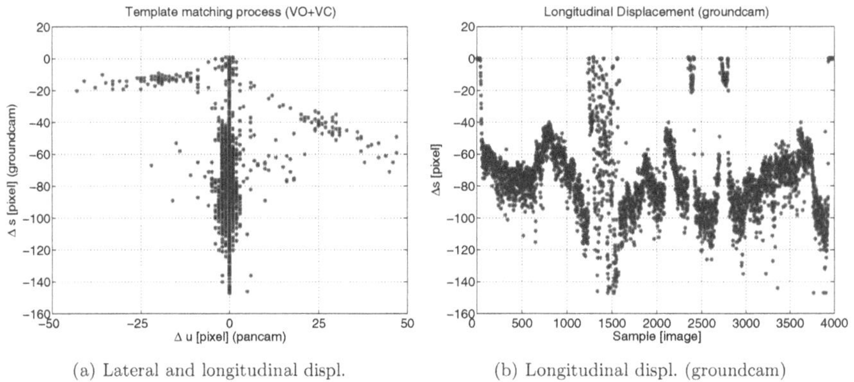

(a) Lateral and longitudinal displ.

(b) Longitudinal displ. (groundcam)

Fig. 3.11 Experiment 2. Template matching using both cameras

Experiment 3. Circular Trajectory

Finally, a circular trajectory was tested in order to check the performance of the proposed localization strategies estimating the robot orientation. The most challenging issue about this experiment is that the robot is always turning and, hence, there are always shadows in the images obtained from the camera pointing at the ground. Furthermore, this circular trajectory highlights the main inconvenience of odometry-based localization techniques, that is, the integration of the orientation from the starting point which lead to inaccurate robot localization for long-range circular trajectories.

In Figure 3.13a, the orientations obtained during the test are shown with respect to the travelled distance. As advised, the typical effect of odometry-based solutions is observed. Notice that the orientation obtained using the odometry-based techniques diverges from the ground-truth (recall that the

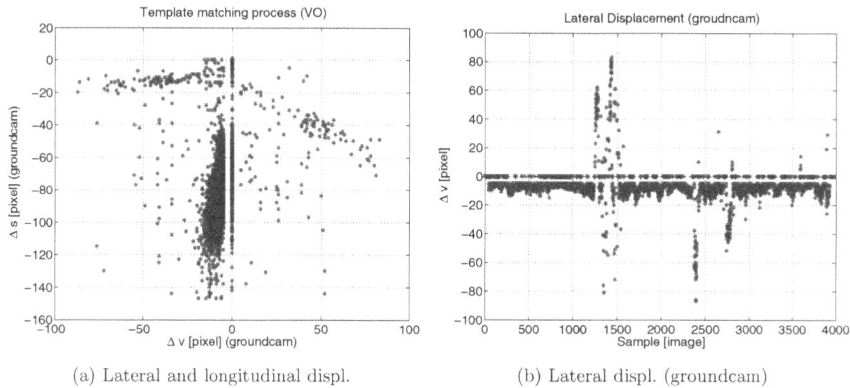

(a) Lateral and longitudinal displ.

(b) Lateral displ. (groundcam)

Fig. 3.12 Experiment 2. Template matching process using Groundcam

experiments were carried out in open-loop). However, it is interesting to remark that acceptable behaviour of the orientation obtained using the visual compass technique. For instance, note that outliers and mismatches highly affect to the estimated orientation using only the camera pointing at the ground. The mean errors between the ground-truth and orientation are: 15.02 [deg] for visual odometry with visual compass, 40.43 [deg] for visual odometry, and 107.84 [deg] for wheel-based odometry.

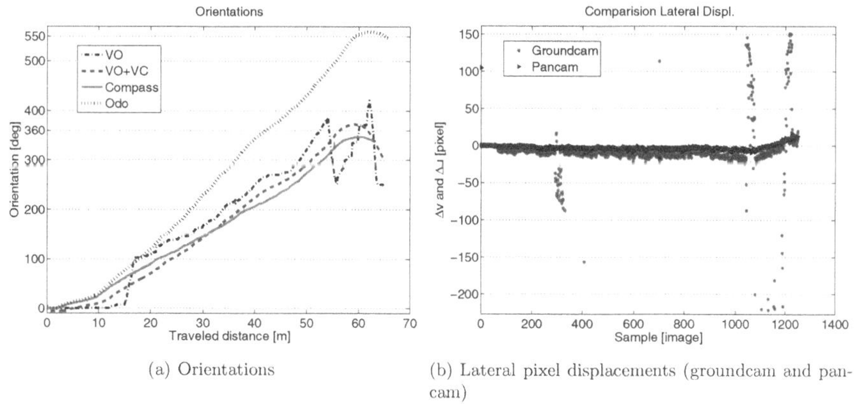

(a) Orientations

(b) Lateral pixel displacements (groundcam and pancam)

Fig. 3.13 Experiment 3. Circular experiment

As observed in Figure 3.13b, the template matching result for the camera pointing at the ground (denoted as "Groundcam") suffers from outliers and false matches, especially at the end of the experiment. This explains the behaviour observed in Figure 3.13a. The images employed by the visual compass approach (labeled as "Pancam") are not affected by shadows and, hence, there are no significant outliers during the matching process.

3.4 Conclusions

In this chapter, the application of visual odometry based on template matching to off-road mobile robots is presented. Standard visual odometry has been improved using the visual compass method to calculate the robot orientation. This strategy has been implemented using two consumer-grade monocular cameras. Physical experiments have confirmed the appropriate behaviour of the proposed scheme. As checked in these experiments, the low computation time ($< 0.2[s]$) will permit to use this localization technique in a robot navigation architecture with a low computation burden. The point related to reduce the size of the search area during the matching process can be considered as an incipient approach to decrease the computation time. However, a best improvement consists in to use the robot motion to reduce accordingly the image size. A multi-objective problem (success of matching process, reduction of template size, reduction of image size) for this issue will be also considered for future works. The problem with shadows and, hence, false matches, constitutes the most important shortcoming of the vision-based strategies. In this book, a deep analysis has been carried out regarding solving this issue. It has been selected a proper downward camera position and a threshold filter. Nevertheless, three ways to improve both solutions will be studied: (i) new positions for the camera pointing at the ground, for instance, just below the robot as in [27]; (ii) new filters and new matching procedures that minimize the effect of shadows and do not increase the computation burden. The problem with blur effect will be treated considering a better camera and other mounting devices; (iii) combining current visual-odometry data with other "localization" techniques through a probabilistic approach.

In conclusion, from physical experiments, the visual odometry strategy is considered as a satisfactory localization technique for middle-size trajectories and for uniform lighting conditions. An important point about this option is that two consumer-grade cameras can replace expensive sensors, such as the Doppler radar, the magnetic compass and the encoders. This point can be really advisable for future commercial or low-cost projects. Additionally, visual information can also be used to estimate slip, by using the methodology proposed in Chapter 2.

Chapter 4
Adaptive Motion Controllers for Tracked Robots

This chapter presents two slip compensation control strategies. The first strategy updates feedback gains taking into account the estimated slip. The second one is an adaptive control strategy formulated using LMI. In the latest one, asymptotic stability, and state and inputs constraints fulfillment is ensured. The reported results confirm that the slip compensation motion controllers are suitable for tracked robots and improve the performance of similar schemes with no slip compensation.

4.1 Introduction

Mobile robots must have effective motion controllers that handle the properties of the surrounding environment. In this context, motion control is defined as a layer in the navigation architecture that generates the proper control actions to successfully steer the mobile robot through a desired trajectory. Particularly, motion controllers should be designed taking into account the robot-terrain interactions, such as the slip phenomena. In this sense, the motion control approaches for mobile robots in off-road conditions are usually achieved from two points of view.

On the one hand, some approaches try to avoid slip generating control signals such that the soil never fails[1]. For instance, in [71], a torque controller is presented. The basic idea of the control algorithm is to set torques for appropriately maximizing traction. In order to measure these torques, novel wheel ground contact angle sensors are built. A wheel-ground contact force control law is discussed in [57]. Authors define an optimization criterium to maximize traction and to minimize power consumption, in such a way that soil failure is avoided. A model-predictive trajectory-tracking controller is

[1] To avoid terrain failure and the resulting wheel/track slip, the control algorithm should seek to maximize the tractive force in a way which that force does not violate the terrain traction constraints [58].

R. González, F. Rodríguez, and J.L. Guzmán, *Autonomous Tracked Robots in Planar Off-Road Conditions,* Studies in Systems, Decision and Control 6,
DOI: 10.1007/978-3-319-06038-5_4,

presented in [67]. The proposed scheme includes velocity and acceleration constraints to prevent the mobile robot from slip. In the work [2], authors propose an approach for slip prediction from a distance for wheeled robots using visual information. Using that information, the robot can avoid entering into slip terrains.

On the other hand, some control strategies adapt the control signals depending on the estimated slip, and, hence, compensate its effect. This is the most general case, since usually slip can never be fully avoided. For instance, in [22], [78], a path tracking controller is designed. That controller uses an extended kinematic model that permits to take into account wheel slip online. Additionally, a predictive algorithm is developed in order to address delays induced by steering actuators and also compensating for transient overshoots in curves. The work [88] proposes a kinematic approach to improve motion control and pose estimation of a TMR. In this case, the robot location is estimated determining the ICR positions, which depend on the track-soil interactions. In the work [62], two different control approaches to compensate for three types of slip, namely, the vehicle sideslip, longitudinal and lateral slips are presented. One is a model-based feedforward control and the second one is a sensor-based feedback control. The paper [50] shows a path following controller that compensates for slip modifying appropriately the velocities sent to the wheels. In this case, slip is calculated as the difference between the Kalman filter position estimate and the kinematic estimate. The Kalman filter combines data from an IMU sensor and visual odometry.

Therefore, from the previous analysis, it can be concluded that the control strategies dealing with slip are grouped into those avoiding the slip phenomenon and those others trying to compensate it. The control laws that avoid the slip by limiting the velocity/acceleration of wheels or tracks [67, 87, 109], can become unsuitable in practice. The reason is that off-road terrains are intrinsically loose, producing a non-controllable slip, that is, the robot will slip although velocity and acceleration are limited. Generally, in the case of the approaches that try to avoid slip, the control actions come from complex dynamic models with numerous parameters that are difficult to be measured online. Furthermore, the control actions can become too conservative.

Hence, this chapter presents two new control techniques for the second approach where the slip is estimated and used in the control law to compensate its effect. The first one is an adaptive linear feedback controller using dynamic feedback linearization and based on a modification of the well-known linear feedback controller described in [21, 64]. The second one is an adaptive control strategy formulated using Linear Matrix Inequalities (LMI). Both approaches have been tested using a tracked mobile robot.

Note that, LMI have an extensive application in the field of automatic control, involving robust and optimal control strategies [15, 49, 69]. LMI-based solutions have also been satisfactorily applied in mobile robotics. For instance, in [11], a feedback path controller for an articulated mining vehicle based on LMI techniques to guarantee stability of the closed-loop system is proposed. In [135], a robust tracking problem of a WMR subject to non-holonomic constraints and input constraints is discussed. In the framework of LMI, the suggested tracking scheme is formulated as an online controller, which is obtained solving a constrained H_∞ control law. The work [89] shows a method for motion planning of mobile robots. The free configuration space is decomposed into Delaunay triangles, and an optimum channel from initial to goal configurations is found by solving an LMI system. Backing control of simulated mobile robots with multiple trailers is presented in [121]. LMI are used to solve the problem of finding stable feedback gains and a common Lyapunov function.

This chapter is organized as follows. Section 4.2 concerns with the slip compensation adaptive controller using dynamic feedback linearization. In Section 4.3, the adaptive controller using LMI is described. In Section 4.4, the performance of the proposed control approaches is shown through physical experiments. Finally, Section 4.5 presents some conclusions and future research.

4.2 Slip Compensation Adaptive Controller Using Dynamic Feedback Linearization

Let us consider the well-known linear control law presented in [21], [64]

$$\begin{bmatrix} u_1'(t) \\ u_2'(t) \end{bmatrix} = \begin{bmatrix} -\kappa_1(t) & 0 & 0 \\ 0 & -\kappa_2(t)\text{sign}(v^{rf}(t)) & -\kappa_3(t) \end{bmatrix} \begin{bmatrix} e_x(t) \\ e_y(t) \\ e_\theta(t) \end{bmatrix}, \tag{4.1}$$

where $\kappa_i \in \mathbb{R}$ are the time-dependent feedback gains, and *sign(·)* is the sign operator. Notice that, the trajectory tracking error model (2.22) is used, and that $e = [e_x \quad e_y \quad e_\theta]^T$ is the state vector, and $u' = [u_1' \quad u_2']^T$ is the input vector.

Replacing (4.1) into (2.22), the closed-loop system becomes

$$\dot{e}(t) = A_{cl}'(t)e(t) = \begin{bmatrix} -\kappa_1(t) & \omega^{rf}(t) & 0 \\ -\left(\omega^{rf}(t) - \vartheta(t)\right) & 0 & v^{rf}(t) \\ 0 & -\kappa_2(t)\text{sign}\left(v^{rf}(t)\right) & -\kappa_3(t) \end{bmatrix} e(t). \tag{4.2}$$

The characteristic polynomial of the closed-loop system is given by

$$\det\big(sI_3 - A'_{cl}(t)\big) = \det\left(\begin{bmatrix} s+\kappa_1(t) & -\omega^{rf}(t) & 0 \\ (\omega^{rf}(t)-\vartheta(t)) & s & -v^{rf}(t) \\ 0 & \kappa_2(t)\mathrm{sign}\big(v^{rf}(t)\big) & s+\kappa_3(t)\end{bmatrix}\right),$$

where s is the Laplace-domain variable and $\det(\cdot)$ means the determinant operator. Notice the dependence on time of closed-loop matrix A'_{cl}, it means that equation (4.3) is defined for all the possible values of matrix A'_{cl}. Solving the determinant and choosing $\kappa_1 = \kappa_3$ [21], the characteristic polynomial follows as

$$\det\big(sI_3 - A'_{cl}(t)\big) = \big(s+\kappa_1(t)\big)\Big(s\big(s+\kappa_1(t)\big) + \\ +\kappa_2(t)\mathrm{sign}\big(v^{rf}(t)\big)v^{rf}(t) + \big(\omega^{rf}(t)\big)^2 - \big(\vartheta(t)\omega^{rf}(t)\big)\Big). \tag{4.3}$$

Supposing that the closed-loop system dynamics is a third-order system of the form [21]

$$(s+2\xi\alpha)(s^2+2\xi\alpha s+\alpha^2) = 0, \tag{4.4}$$

where $\xi \in \mathbb{R}^+$ and $\alpha \in \mathbb{R}^+$ are the damping factor and natural frequency, respectively. Now, equaling (4.4) to (4.3), the feedback gains are defined as

$$\kappa_1(t) = 2\xi\Big(\big(\omega^{rf}(t)\big)^2 + \beta\big(v^{rf}(t)\big)^2\Big)^{\frac{1}{2}}, \tag{4.5}$$

$$\kappa_2(t) = \beta \mid v_{ref}(t) \mid + \omega^{rf}(t)\vartheta(t), \tag{4.6}$$

$$\kappa_3(t) = 2\xi\Big(\big(\omega^{rf}(t)\big)^2 + \beta\big(v^{rf}(t)\big)^2\Big)^{\frac{1}{2}}, \tag{4.7}$$

where $\alpha = \Big(\big(\omega^{rf}(t)\big)^2 + \beta\big(v^{rf}(t)\big)^2\Big)^{\frac{1}{2}}$ with $\beta > 0$, $\beta > 0$ and $\xi > 0$ are constants experimentally tuned, and $\mid \cdot \mid$ denotes the element-wise absolute value. Notice that, equations (4.5)-(4.7) differ with respect to those obtained in the original formulation [21] due to the presence of ϑ variable into equation (4.6). In this sense, κ_2 is updated online based on the slip estimation and, thus, compensating its effect.

The main features of this controller are:

- It constitutes a straightforward extension of the well-known linear feedback control law [21, 64, 106], but now, slip explicitly appears in the formulation of the feedback gains. Thus, the control law adapts to the current slip.
- Following the trajectory tracking paradigm, the location (position and orientation) and velocity of the robot are controlled. In the path tracking approach only the steering motion of the robot is handled.

- Efficient real-time execution, since this control strategy demands a really small computation load. This supposes that small sampling rates can be reached.

On the other hand, the main limitations of the proposed control approach are:

- It does not take into account state and input constraints. This means that it can produce control signals that cannot be reached by actuators or it may lead to large errors. Mapping constraint approach should be considered to face this issue.
- Stability is not ensured.

In order to improve these inconveniences, other control strategies have been designed in this book. These control strategies are presented in the following section and in Chapter 5.

4.3 Slip Compensation Adaptive Controller Using an LMI-Based Approach

This section focuses on the synthesis of an adaptive control law that guarantees asymptotic stability subject to state and input constraints and varying dynamics. This controller uses the discrete-time linear time-varying trajectory tracking error model defined in (2.26). The objective is to determine an adaptive control law and a Lyapunov function guaranteeing the asymptotic stability of the closed-loop system. This objective is formulated in terms of an optimization problem, where constraints are formulated using LMI. The motivation to use LMI in the convex optimization problem is that such optimization problem can be solved in polynomial time using powerful algorithms that rapidly compute the global optimum [69]. Once the optimization problem is solved, the definite positive matrix P, which determines the Lyapunov function, is obtained. Furthermore, a set of feedback gains composing the control law are calculated. Finally, these feedback gains are combined online to produce the most proper gain according to the system realization.

The control strategy is designed to satisfy the following specifications:

- Adaptivity. The control has to fulfill the specifications for any admissible realization of the matrix A_γ in model (2.26) (see Remark 3, Chapter 2). For that reason, a feedback gain is designed for any extremal realization of the linear system, such that the quadratic function is a common Lyapunov function. The set of controllers induces a time-varying feedback gain.
- Asymptotic stability. It is achieved by means of a quadratic Lyapunov function, that is, a positive definite function decreasing along the trajectories of the closed-loop system.

- Performance. The quadratic Lyapunov function provides an upper bound on the cost-to-go as close as possible to the optimal Linear Quadratic Regulator (LQR) cost. Recall that, with cost-to-go, the sum of the stage cost from the present to the infinite time is considered. This solution provides a cost-to-go function, which is an overbound of the optimal LQR one. The objective of the optimization problem is to minimize such overbounding function, in order to make the cost-to-go as close as possible to the optimal one, i.e., the LQR cost.
- Input and state constraints fulfillment. This requirement is guaranteed through the determination of an invariant set [12], i.e., a set in which the system state can be confined by the control law. An ellipsoidal invariant set is computed ensuring constraints satisfaction and providing a feasibility region for the controller.
- Performance region. A target set is defined where the performance is considered. The objective is that the controller obtained solving the optimization problem has higher performance inside this set, which is the region of the state space in which the system is confined in practice.
- Efficient real-time execution. Once the feedback gains have been calculated for each extremal system realization, it is sufficient to solve an online linear programming problem in order to obtain the stabilizing control law. This fact supposes that the control strategy detailed here is suitable for mobile robotics applications, where low sampling rates are employed.

The suggested control scheme is summarized in Figure 4.1. First, an offline optimization problem that generates a set of feedback gains is solved, one gain for each extreme realization of the system dynamics. The constraints of that optimization problem are formulated in terms of LMI ensuring asymptotic stability, input and state constraints fulfillment, and location of the performance. Afterwards, a feedback control gain is determined online depending

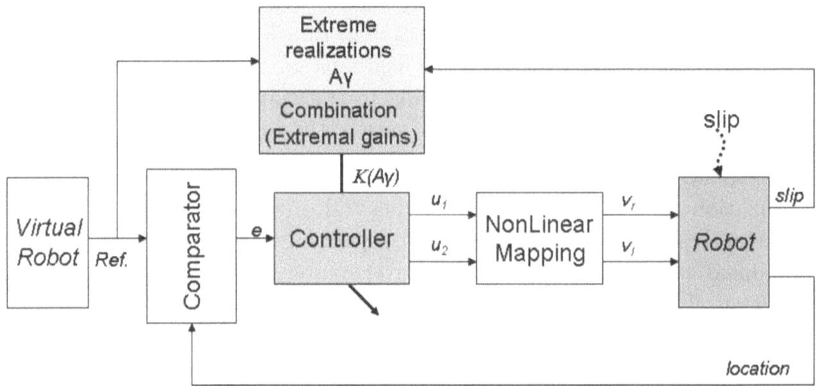

Fig. 4.1 Control scheme for the adaptive control strategy using LMI

on the current system realization. This adaptive control gain is obtained as a convex combination of the previously determined gains. Finally, the constraints mapping procedure addressed in Chapter 2 (Remark 4) is employed to generate the set-points to each track.

4.3.1 Problem Statement

In order to obtain the adaptive control strategy which fulfills previous specifications, the following control law is proposed,

$$u(k) = K(A_\gamma(k))e(k), \tag{4.8}$$

where the feedback gain, $K(A_\gamma(k))$, is obtained as

$$K(A_\gamma(k)) = \varrho_1 K(A_\gamma^1) + \ldots + \varrho_{n_\gamma} K(A_\gamma^{n_\gamma}), \tag{4.9}$$

being $K(A_\gamma^j) \in \mathbb{R}^{2\times 3}$ the obtained offline feedback gains for each extreme realization of the system dynamics (2.26). Notice that the current feedback gain, $K(A_\gamma(k))$, is solved as a convex combination of a set of feedback gains $K(A_\gamma^j)$. The problem to determine the vector of coefficients ϱ is a well-known Linear Programming feasibility problem (LP), which is related to

$$A_\gamma(k) = \sum_{j=1}^{n_\gamma} \varrho_j A_\gamma^j, \qquad \sum_{j=1}^{n_\gamma} \varrho_j = 1, \quad \varrho_j \geq 0, \quad \forall j = 1, \ldots, n_\gamma, \tag{4.10}$$

where A_γ is the current matrix (online realization), and A_γ^j is the j-th extreme realization of the set $\mathcal{A}^\gamma$.

From previous discussion, the closed-loop system becomes

$$e(k+1) = \bar{A}_\gamma(k)e(k) = \big(A_\gamma(k) + B_d K(A_\gamma(k))\big)e(k). \tag{4.11}$$

When dealing with the problem of determining asymptotically stable controllers, a classical way to proceed is to look for a Lyapunov function determined by a definite positive matrix $P > 0$, i.e., $V(e) = e^T P e$, such that

$$V(e(k+1)) - V(e(k)) \leq 0, \tag{4.12}$$

for all $e \neq 0$ [15, 69].

In this book, asymptotic stability and performance objectives for the closed-loop system (4.11) are defined as:

- The function $V(e) = e^T \mu P e$ is a local Lyapunov function for the system inside $\epsilon(P) = \{e : e^T P e \leq 1\}$ ensuring stability. The meaning of variable μ is explained subsequently.

- The set $\epsilon(P) = \{e : e^T P e \leq 1\}$ is an invariant set for the closed-loop system (4.11) satisfying input and state constraints. Note that $\epsilon(P)$ is a level set of a Lyapunov function [13].
- The function $V(e)$ is a upper bound of the cost-to-go, and of the cost of the LQR i.e.,

$$V\big(e(0)\big) \geq \min_{u_{[0,\infty)}} \sum_{k=0}^{\infty} e^T(k) Q_L e(k) + u(k)^T R_L u(k), \tag{4.13}$$

  where $u_{[0,\infty)}$ denotes the infinite sequence of $u(k)$ $\forall e \in \epsilon(P)$, and $Q_L > 0$ and $R_L > 0$ are symmetric matrices weighting the state and input signals.

In general, the invariant ellipsoid and the Lyapunov function are both determined by P. A further degree of freedom introducing a scaling factor, μ, is added in the definition of the Lyapunov function leading to $V(e) = e^T \mu P e, \mu \in \mathbb{R}^+$. The value of μ provides an upper bound of the cost-to-go valid within a performance region. Hence, conceptually, minimizing μ implies maximizing the performance in a region of interest. Roughly speaking, two opposite objectives are considered. Small values of μ lead to low bounds of the cost-to-go, and this means high performance over a narrow region. So, in order to avoid such "narrow region", a target region, Ψ, is included, as a constraint in the minimization problem, in which the system is confined at beginning of an experiment. That is, the initial error between the robot and the reference must be within Ψ.

Then, from previous requirements the following optimization problem can be proposed

$$\begin{array}{cc} \displaystyle\min_{P>0,\ \mu,\ K(A_\gamma^j)\,\forall\gamma^j} & \mu \\ \text{subject to} & L^*\ \forall\gamma^j, \end{array} \tag{4.14}$$

where L^* is defined by the following inequality

$$\begin{aligned} L^* = e^T\big((\bar{A}_\gamma^j)^T \mu P \bar{A}_\gamma^j\big)e - e^T(\mu P)e \leq \\ \leq -e^T\Big(Q_L + \big(K(A_\gamma^j)\big)^T R_L K(A_\gamma^j)\Big)e\,\forall e \in \mathbb{R}^3. \end{aligned} \tag{4.15}$$

Notice that, in this minimization problem all the required conditions are imposed only at the extremal values of the polytopic set Γ (see Remark 3, Chapter 2), i.e., at the n_γ vertices of Γ. In this case, $n_\gamma = 2^4$, since as shown in Assumption 1, system matrix A_γ is determined by the admissible realization of four variables $(v_r^{rf}, v_l^{rf}, i_r, i_l)$. The fulfillment of such conditions at the vertices yields the satisfaction at any point in Γ, as stated in Property 1, that will be presented in Subsection 4.3.6.

In the following subsection, the inequality regarding asymptotic stability and performance (4.15) will be translated to LMI form in order to address the optimization problem (Subsection 4.3.2). Furthermore, input and state constraints (Subsections 4.3.3 and 4.3.4), and a constraint dealing with the performance region will be formulated in terms of LMI (Subsection 4.3.5). Finally, the minimization problem is completely stated (Subsection 4.3.6).

4.3.2 Asymptotic Stability and Performance

Taking into account asymptotic stability constraint (4.12) and the performance objective (4.13), the following inequality is considered

$$e^T\Big((A_\gamma^j + B_d K(A_\gamma^j))^T \mu P (A_\gamma^j + B_d K(A_\gamma^j))\Big)e - e^T(\mu P)e \leq \\ \leq -e^T\Big(Q_L + (K(A_\gamma^j))^T R_L K(A_\gamma^j)\Big)e, \quad \forall e \in \mathbb{R}^3, \tag{4.16}$$

for every vertex γ^j of Γ, with $j = 1, \ldots, n_\gamma$. Notice that, for notational convenience, the dependence on time k is omitted.

Now, taking into account (4.11), equation (4.16) follows as

$$e^T((\bar{A}_\gamma^j)^T \mu P \bar{A}_\gamma^j)e - e^T(\mu P)e \leq \\ \leq -e^T\Big(Q_L + (K(A_\gamma^j))^T R_L K(A_\gamma^j)\Big)e, \forall e \in \mathbb{R}^3. \tag{4.17}$$

The previous inequality is equivalent to the following LMI [15]

$$(\bar{A}_\gamma^j)^T \mu P \bar{A}_\gamma^j - \mu P \leq -Q_L - (K(A_\gamma^j))^T R_L K(A_\gamma^j). \tag{4.18}$$

Using the Schur complement and replacing $S = P^{-1}$, $Y_\gamma^j = K(A_\gamma^j)P^{-1}$, and $\bar{A}_\gamma^j = A_\gamma^j + B_d K(A_\gamma^j)$, the LMI follows as [40]

$$\begin{bmatrix} S & S(A_\gamma^j)^T + (Y_\gamma^j)^T B_d^T & S Q_L^{\frac{1}{2}} & (Y_\gamma^j)^T R_L^{\frac{1}{2}} \\ A_\gamma^j S + B_d Y_\gamma^j & S & 0 & 0 \\ Q_L^{\frac{1}{2}} S & 0 & \mu I & 0 \\ R_L^{\frac{1}{2}} Y_\gamma^j & 0 & 0 & \mu I \end{bmatrix} \geq 0, \tag{4.19}$$

for every vertex γ^j of Γ, with $j = 1, \ldots, n_\gamma$.

4.3.3 Input Constraints

Physical limitations in the actuators must be included to the previous optimization problem (4.14) in the LMI form. Knowing that $v_r \leq v_r^M$ and $v_l \leq v_l^M$, such restriction in terms of the virtual control inputs is given by[2]

[2] Notice that the tracks are considering to always move forward. Nevertheless, in practice, the tracks can also move backward. However, these negative velocities are rarely reached, since the reference virtual robot always moves forward. For that reason, no lower bounds have to be added to the optimization problem.

$$\frac{v_r^{rf} - u_1 - \frac{b}{2}u_2}{1 - i_r} \leq v_r^M, \tag{4.20}$$

for the right track, the case for v_l is obtained in a similar way.

Now, the previous constraint is formulated for each extreme realization of the system dynamics, that is, $n_\gamma = 2^4$, since as shown in Assumption 1, system matrix A_γ is determined by the admissible realization of the four variables $(v_r^{rf}, v_l^{rf}, i_r, i_l)$. In this way, equation (4.20) follows as

$$C_r^j u + d_r^j \leq v_r^M, \tag{4.21}$$

for every vertex γ^j of Γ, with $j = 1, \dots, n_\gamma$, where $C_r^j = \left[\frac{-1}{1-i_r} \;\; \frac{-b}{2(1-i_r)} \right]$, $d_r^j = \frac{v_r^{rf}}{1-i_r}$, and i_r and v_r^{rf} are those related to the particular extremal realization γ^j of the parameter. Recall that $u = [u_1 \quad u_2]^T$.

Defining $\eta_r^j = v_r^M - d_r^j$ and replacing $u = K(A_\gamma^j)e$, the previous inequality becomes

$$C_r^j K(A_\gamma^j)e \leq \eta_r^j, \quad \forall e \in \epsilon(P). \tag{4.22}$$

Considering the following problem to determine the maximum value of a linear function with ellipsoidal constraints, i.e.,

$$\begin{aligned} a^* = \max_e \quad & C_r^j K(A_\gamma^j)e \\ \text{subject to} \quad & e^T P e \leq 1, \end{aligned} \tag{4.23}$$

the solution to the previous maximization problem is given by [15]

$$a^* = \sqrt{C_r^j K(A_\gamma^j) P^{-1} \left(K(A_\gamma^j)\right)^T (C_r^j)^T}. \tag{4.24}$$

Hence, a necessary and sufficient condition for (4.22) to be fulfilled is that

$$C_r^j K(A_\gamma^j) P^{-1} \left(K(A_\gamma^j)\right)^T (C_r^j)^T \leq (\eta_r^j)^2. \tag{4.25}$$

Notice that there is a quadratic term $(\eta_r^j)^2$ depending on γ^j. In order to ensure the convexity properties of LMI, this terms is replaced by the upper bound $\bar{\eta}_r = v_r^M - \frac{v_r^{rf,M}}{1-i_r^M}$. Thus, it follows

$$C_r^j K(A_\gamma^j) P^{-1} \left(K(A_\gamma^j)\right)^T (C_r^j)^T \leq (\bar{\eta}_r)^2. \tag{4.26}$$

Then, applying the Schur complement, it becomes

$$\begin{bmatrix} (\bar{\eta}_r)^2 & C_r^j K(A_\gamma^j) \\ \left(K(A_\gamma^j)\right)^T (C_r^j)^T & P \end{bmatrix} \geq 0. \tag{4.27}$$

Finally, previous equation is pre- and post-multiplied by

$$\begin{bmatrix} I & 0 \\ 0 & P^{-1} \end{bmatrix}, \tag{4.28}$$

resulting

$$\begin{bmatrix} (\bar{\eta}_r)^2 & C_r^j K(A_\gamma^j) P^{-1} \\ P^{-1}\left(K(A_\gamma^j)\right)^T (C_r^j)^T & P^{-1} \end{bmatrix} \geq 0, \tag{4.29}$$

now, replacing $S = P^{-1}$ and $Y_\gamma^j = K(A_\gamma^j)P^{-1}$, the input constraints result in the following LMI

$$\begin{bmatrix} (\bar{\eta}_r)^2 & C_r^j Y_\gamma^j \\ (Y_\gamma^j)^T (C_r^j)^T & S \end{bmatrix} \geq 0, \tag{4.30}$$

for every vertex γ^j of Γ, with $j = 1, \ldots, n_\gamma$.

The LMI for the left track is obtained in a similar way, resulting

$$\begin{bmatrix} (\bar{\eta}_l)^2 & C_l^j Y_\gamma^j \\ (Y_\gamma^j)^T (C_l^j)^T & S \end{bmatrix} \geq 0, \tag{4.31}$$

for every vertex γ^j of Γ, with $j = 1, \ldots, n_\gamma$, and where $C_l^j = \begin{bmatrix} \frac{1}{i_l - 1} & \frac{-b}{2(1-i_l)} \end{bmatrix}$ and $\bar{\eta}_l = v_l^M + \frac{v_l^{rf,m}}{1-i_r^m}$.

4.3.4 State Constraints

In addition to the input constraints, it must be imposed that the invariant set is contained in the admissible state space region. This guarantees no state constraints violations, provided that the initial state is confined in $\epsilon(P)$. Using (2.26), state constraints are formulated as

$$|\, e \,| \leq [e_x^M \quad e_y^M \quad e_\theta^M]^T, \quad \forall e \in \epsilon(P), \tag{4.32}$$

then, it can be written as

$$\begin{aligned} |\, f_1 e \,| &\leq e_x^M, \quad \forall e \in \epsilon(P), \\ |\, f_2 e \,| &\leq e_y^M, \quad \forall e \in \epsilon(P), \\ |\, f_3 e \,| &\leq e_\theta^M, \quad \forall e \in \epsilon(P), \end{aligned} \tag{4.33}$$

where $f_1 = [1 \;\, 0 \;\, 0]$, $f_2 = [0 \;\, 1 \;\, 0]$, $f_3 = [0 \;\, 0 \;\, 1]$. Considering the following problem to determine the maximum value of a linear function with ellipsoidal constraints, i.e.,

$$\begin{aligned} b^* = \max_{e} \quad & |(f_i e)| \\ \text{subject to} \quad & e^T P e \leq 1, \end{aligned} \tag{4.34}$$

for $i = 1, 2, 3$. The solution to the previous maximization problem is given by [15]

$$b^* = \sqrt{f_i P^{-1} f_i^T}. \tag{4.35}$$

Hence, defining $S = P^{-1}$, the state constraints are formulated in terms of LMI as

$$\begin{aligned} f_1 S f_1^T &\leq (e_x^M)^2, \\ f_2 S f_2^T &\leq (e_y^M)^2, \\ f_3 S f_3^T &\leq (e_\theta^M)^2. \end{aligned} \tag{4.36}$$

Notice that due to symmetry of LMI, this equation is also achieved for the lower bounds.

4.3.5 Performance Region

In this subsection, the target region, Ψ, is established. Recall from previous discussion that Ψ can be considered as a performance region, in which the system evolves mostly. The objective is that the controller obtained solving LMI has higher performance inside this set. The parameter $\psi_e^j \in \Psi$ is defined as the j-th vertex of Ψ. The set Ψ is the parallelotope in the state space determined by the intervals of interest, i.e., $\Psi = \{e : \psi_e^m \leq e \leq \psi_e^M\}$, and imposing $\Psi \subseteq \epsilon(P)$, it leads to

$$(\psi_e^j)^T P (\psi_e^j) \leq 1, \tag{4.37}$$

which is equivalent to

$$1 - (\psi_e^j)^T P (\psi_e^j) \geq 0. \tag{4.38}$$

Then, applying the Schur complement, it becomes

$$\begin{bmatrix} 1 & (\psi_e^j)^T \\ \psi_e^j & S \end{bmatrix} \geq 0. \tag{4.39}$$

4.3.6 Final Optimization Problem

In conclusion from previous subsections, the offline design process is aimed to obtain a family of control gains fulfilling the LMI constraints and such that the resulting Lyapunov function is induced minimizing with respect to μ, that is,

$$\min_{S>0,\ \mu,\ Y_\gamma^j\ \forall\gamma^j} \quad \mu$$
subject to

$$\begin{bmatrix} S & S(A_\gamma^j)^T + (Y_\gamma^j)^T B_d^T & SQ_L^{\frac{1}{2}} & (Y_\gamma^j)^T R_L^{\frac{1}{2}} \\ A_\gamma^j S + B_d Y_\gamma^j & S & 0 & 0 \\ Q_L^{\frac{1}{2}} S & 0 & \mu I & 0 \\ R_L^{\frac{1}{2}} Y_\gamma^j & 0 & 0 & \mu I \end{bmatrix} \geq 0, \quad \forall\gamma^j,$$

$$\begin{bmatrix} (\bar{\eta}_r)^2 & C_r^j Y_\gamma^j \\ (Y_\gamma^j)^T (C_r^j)^T & S \end{bmatrix} \geq 0, \quad \forall\gamma^j,$$

$$\begin{bmatrix} (\bar{\eta}_l)^2 & C_l^j Y_\gamma^j \\ (Y_\gamma^j)^T (C_l^j)^T & S \end{bmatrix} \geq 0, \quad \forall\gamma^j,$$

$$f_1 S f_1^T \leq (e_x^M)^2,$$

$$f_2 S f_2^T \leq (e_y^M)^2,$$

$$f_3 S f_3^T \leq (e_\theta^M)^2,$$

$$\begin{bmatrix} 1 & (\psi_e^j)^T \\ \psi_e^j & S \end{bmatrix} \geq 0, \quad \forall\psi_e^j.$$

Recall that the solution to this optimization problem gives a set of feedback gains, one gain for each extreme realization of the system dynamics (2.26). Online computation is devoted to solve an LP problem to produce the current feedback gain as a convex combination of these gains obtained offline (4.9).

As an example, Figure 4.2 plots the invariant ellipsoid obtained solving (4.40) using the LMI toolbox [35] and MPT toolbox [72] both for Matlab® suite, and using the values shown in Table 4.1. The small box inside the ellipsoid depicts the set Ψ. As expected, the ellipsoid is constrained by the performance region. In addition, input and state constraints are also drawn. Recall that, once the feedback law is defined, the input constraints are projected into the state space (see equation (4.22)). This is noticed in the top right cuts of the outer box in Figure 4.2.

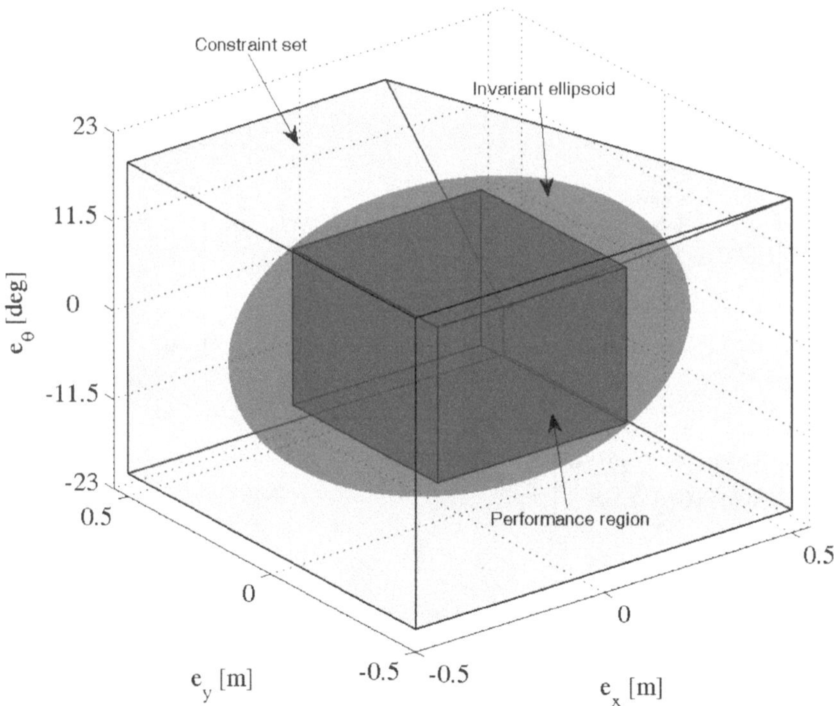

Fig. 4.2 Ellipsoidal invariant set within state constraints set and outwith performance region

Table 4.1 Input and state constraints and performance region for experiments

	Max	Min	Units
Input			
v_r	2	0	[m/s]
v_l	2	0	[m/s]
State			
e_x	0.5	-0.5	[m]
e_y	0.5	-0.5	[m]
e_θ	20	-20	[deg]
Performance region			
ψ_{e_x}	0.25	-0.25	[m]
ψ_{e_y}	0.25	-0.25	[m]
ψ_{e_θ}	10	-10	[deg]

The following property shows that the adaptive control law (4.8) ensures the specifications given at the beginning of this section for any $A_\gamma \in \mathcal{A}^\gamma$.

Property 1. Suppose that Assumption 1 holds (see Chapter 2, Section 2.2). Consider the discrete-time linear time-varying system (2.26) with constraints $e(k) \in E$, $u(k) \in U$ (2.34). The adaptive control law defined in (4.8) is such that, for each $\gamma \in \Gamma$:

- The function $V(e) = e^T \mu P e$ is a local Lyapunov function for the system inside $\epsilon(P) = \{e : e^T P e \leq 1\}$ ensuring stability.
- The set $\epsilon(P) = \{e : e^T P e \leq 1\}$ is an invariant set for the closed-loop system satisfying input and state constraints.
- The function $V(e)$ is a upper bound of the cost-to-go, and of the cost of the LQR, i.e.,

$$V\big(e(0)\big) \geq \min_{u_{[0,\infty)}} \sum_{k=0}^{\infty} e^T(k) Q_L e(k) + u(k)^T R_L u(k), \tag{4.40}$$

where $u_{[0,\infty)}$ denotes the infinite sequence of $u(k)$ $\forall e \in \epsilon(P)$.

Proof:
In the following, it is assumed that $\varrho = [\varrho_1, \cdots, \varrho_{n_\gamma}]$ is obtained such that A_γ is a convex combination of A_γ^j, with γ^j being the n_γ vertices of Γ as in (4.16) and $K(A_\gamma)$ is determined by (4.9). The three statements of the property are proved. Let us start proving the third point, since the first one is a direct consequence of it. Note that, for notational convenience, the dependence on time k is omitted.

To prove the third point, it must be shown that inequality

$$\begin{aligned} e^T\Big(\big(A_\gamma + B_d K(A_\gamma)\big)^T \mu P \big(A_\gamma + B_d K(A_\gamma)\big)\Big)e - e^T(\mu P)e \leq \\ \leq -e^T\big(Q_L + K(A_\gamma)^T R_L K(A_\gamma)\big)e, \quad \forall e \in \mathbb{R}^3, \end{aligned} \tag{4.41}$$

is satisfied for any $\gamma \in \Gamma$, i.e., if it is satisfied at the vertices γ^j, with $j = 1, \ldots, n_\gamma$. To simplify the notation, the following matrix is defined

$$M_\gamma = \begin{bmatrix} S & S(A_\gamma)^T + (Y_\gamma)^T B_d^T & S Q_L^{\frac{1}{2}} & (Y_\gamma)^T R_L^{\frac{1}{2}} \\ (A_\gamma)S + B_d(Y_\gamma) & S & 0 & 0 \\ Q_L^{\frac{1}{2}} S & 0 & \mu I & 0 \\ R_L^{\frac{1}{2}} Y_\gamma & 0 & 0 & \mu I \end{bmatrix}, \tag{4.42}$$

where $S = P^{-1}$ and $Y_\gamma = K(A_\gamma)P^{-1}$. Notice that, the dependence on the parameter γ is explicit, and that $M_\gamma = \sum_{i=1}^{n_\gamma} \varrho^j M_{\gamma^j}$. Since, as illustrated in Subsection 4.3.2, condition (4.41) is equivalent to $M_\gamma \geq 0$, and $\varrho_j \geq 0$ from (4.10), this leads to

$$M_\gamma = \sum_{i=1}^{n_\gamma} \varrho^j M_{\gamma^j} \geq 0, \tag{4.43}$$

what means that (4.41) is fulfilled. To prove that $V(e)$ is a Lyapunov function for the closed-loop system regardless of the realization of parameter γ, it must be shown that for any $\gamma \in \Gamma$, condition $V(e(k+1)) - V(e(k)) < 0$ is satisfied, that is,

$$e^T\Big(\big(A_\gamma + B_d K(A_\gamma)\big)^T \mu P\big(A_\gamma + B_d K(A_\gamma)\big)\Big)e - e^T(\mu P)e < 0, \quad \forall e \in \mathbb{R}^3,$$

since $V(e)$ is positive definite. From $Q_L > 0$, $R_L > 0$ and (4.41), the condition follows. Furthermore, since it is proved for the entire space, $V(e)$ is, in particular, a Lyapunov function in the ellipsoid centered in the origin. Finally, the set $\epsilon(P)$ is an invariant set since it is the level set of a Lyapunov function (see [13]), and it fulfills the state constraints by construction. For what concern the input constraints, it is obtained that

$$C_\gamma K(A_\gamma)e = \sum_{j=1}^{n_\gamma} \varrho^j C_{\gamma^j} K(A_\gamma^j)e \leq \sum_{i=1}^{n_\gamma} \varrho^j \bar{\eta} = \bar{\eta} \quad \forall e \in \epsilon(P), \tag{4.44}$$

is satisfied by convexity of Γ and fulfillment of input constraints at its vertices. ■

4.4 Results

This section analyzes the performance of the two proposed slip compensation control strategies, the slip compensation adaptive controller using dynamic feedback linearization (Section 4.2) and the adaptive control law using LMI (Section 4.3). Additionally, the original formulation of the linear feedback controller, detailed in [21] and [64], has also been implemented for comparison purposes. From now on, the slip compensation adaptive controller using LMI is referred to as LMI-based controller, the slip compensation adaptive controller using dynamic feedback linearization is referred to as FL controller, and the original linear feedback controller as OC controller.

Some of the configuration parameters employed for physical experiments are summarized in Table 4.1. The rest of parameters are: $b = 0.5[m]$, reference velocities are restricted to $\{v_r^{rf}, v_l^{rf} \in [0.1, 1.4][m/s]\}$ and slips are restricted to $\{i_r, i_l \in [0, 25][\%]\}$. The own parameters of the LMI-based control law are $Q_L = diag([1 \quad 1 \quad 0.1])$ and $R_L = 10I_2$. The parameters of the linear feedback controllers are set to $\beta = 1$ and $\xi = 0.6$ in order to reach a soft overdamped closed-loop behaviour. Notice that Q_L and R_L matrices have been tuned empirically to achieve a satisfactory performance. The initial location of the mobile robot is always $[0 \quad 0 \quad 0]^T$. A sampling period of $T_s = 0.35[s]$. This sampling period has been selected according to the desired specifications of the closed-loop system.

4.4.1 Experimental Results

This subsection describes two physical experiments to check the performance of the three compared control strategies. For that purpose, the mobile robot *Fitorobot* was employed. Before discussing experiments, some considerations about practical issues must be pointed out:

- Recall, from the introductory chapter, that a typical four-layer navigation architecture has been designed and implemented in order to attempt physical experiments (see Figure 1.3).
- The visual odometry (with visual compass) approach has been used for localize the robot. The slip of both tracks were estimated comparing the actual robot velocity, obtained using visual odometry, and the theoretical velocities of the tracks, obtained using incremental encoders. In this case, the ground-truth is determined using the DGPS (position) and the magnetic compass (orientation) previously employed in Chapters 2 and 3. The DGPS has an accuracy of 0.20 [m] and the magnetic compass of 0.1 [deg].
- The tests were carried out on an off-road gravel terrain, similar to that presented in Figure 2.8 (Chapter 2, Section 2.7). Although this is not an ideal site for the experiments, since it is a partially bumpy terrain, the results presented here are illustrative of the performance of the compared controllers.
- The experiments were carried out with sunlit conditions, and thus some shadows that can lead to false matches in the visual odometry algorithm were observed. Nevertheless, these false matches were minimized due to the two approaches explained in Chapter 3 (Section 3.3.1): appropriate downward camera position and threshold filter.
- Although many experiments have been carried out, in this case a circular trajectory and a U-shaped trajectory have been selected. In the former the robot moved at middle/high velocities and in the second one it moved at low velocities. Notice that similar trajectories to these selected here are usually employed in off-road mobile robotics, see for instance [76, 79, 104, 115, 137].

Experiment 1. Circular Trajectory

In this first experiment, a circular reference trajectory was tested. The motivation by selecting this trajectory is because it is a closed trajectory, what helps in a better comparison of the controllers, since the robot has to come back to the starting point. Furthermore, the reference track velocities were 0.6 [m/s] for the right track and 0.5 [m/s] for the left track. For the testbed, these can be considered as middle/high velocities. In this first selected experiment, the total travelled distance was close to 18 [m].

Figure 4.3a shows the reference trajectory and the followed trajectories using the three control strategies. Figure 4.3b shows the reference trajectory, the trajectory followed using the LMI-based controller, and the ground-truth

(DGPS). In the former plot, it is possible to observe that the LMI-based controller achieves the best behaviour. However, as expected, it does not fix the reference trajectory due to the disturbances and inconveniences of physical experiments. Nevertheless, all the compared controllers achieves a proper result and the mobile robot comes back to the initial point of the circular trajectory. An interesting conclusion is obtained from Figure 4.3b. As remarked at the beginning of this subsection (considerations about practical issues), the trajectory obtained using the LMI-based controller does not fix the ground-truth (DGPS). In this particular experiment, the maximum lateral deviation is 0.52 [m], and the mean lateral deviation is 0.12 [m], what means 0.67 [%] of the total travelled distance.

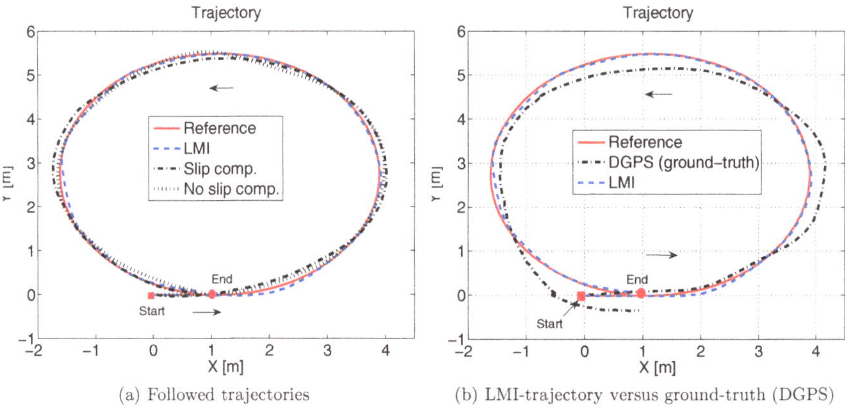

(a) Followed trajectories

(b) LMI-trajectory versus ground-truth (DGPS)

Fig. 4.3 Experiment 1. Followed trajectories during the experiment

In Figure 4.4a, the orientations are shown. The proper behaviour of the LMI-based controller with few oscillations can be observed. Figure 4.4b plots the reference orientation, the orientation obtained using the LMI-based controller, and the ground-truth (magnetic compass). As expected, from the discussion about Figure 4.3b, the orientation obtained using the LMI-based controller follows the ground-truth with small deviations. Particularly, the mean deviation is 7.62 [deg] with standard deviation 4.12 [deg].

Figure 4.5 shows the slip estimations for each track. Recall that the slip is obtained as the relation between the actual robot velocity (visual odometry) and the theoretical velocities of the tracks (incremental encoders). In this particular experiment, the tracks move at different velocities. For that reason, the obtained slip estimations are different. Notice that this fact can lead to a mistake since the same actual velocity is being supposed for both tracks. In order to minimize this mistake related to circular trajectories, the mean value of both slip estimations is considered. It is assumed that this solution holds for the tested range of velocities. In further works, two independent cameras will be employed to estimate the actual velocity of each track. The mean slip value is 4.41 [%] that confirms the results presented in Chapter 2 about the mean slip value on a gravel soil.

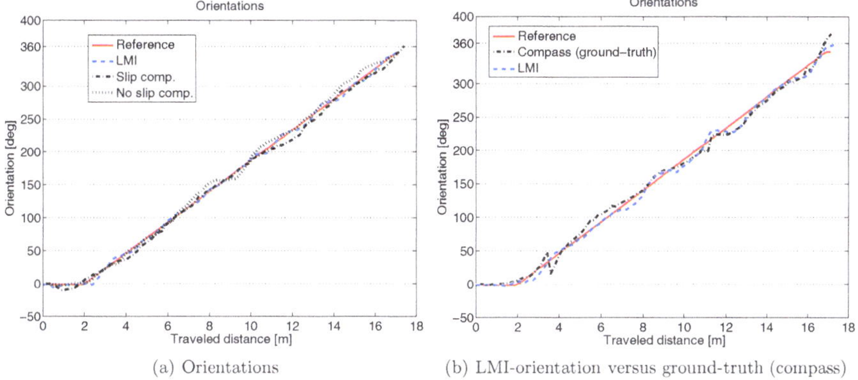

(a) Orientations (b) LMI-orientation versus ground-truth (compass)

Fig. 4.4 Experiment 1. Orientations

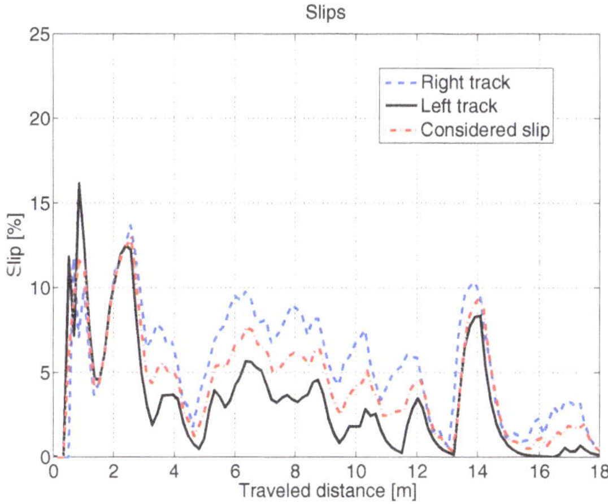

Fig. 4.5 Experiment 1. Slip estimations for the LMI-based controller (comparing the robot velocity using visual odometry and track velocities using encoders)

The errors between the reference trajectory and those steered by the compared controllers are displayed in Figure 4.6. From these plots, it is noticed that the best behaviour is obtained for the LMI-based controller. The suggested FL controller also achieves a satisfactory result in comparison to the OC controller. A positive longitudinal error is observed. This means that the real robot is pursuing the reference robot. Furthermore, from these plots, it can be observed a maximum longitudinal error of the linear feedback controllers close to 0.50 [m], which can be considered as unappropriate. In relation to the lateral error, the OC controller shows a noticeable oscillatory behaviour with maximum values of 0.15 [m]. In this case, it means that the robot can crash with obstacles in a narrow workspace. The mean longitudinal errors are: 0.05 [m] with standard deviation 0.06 [m] for the LMI-based controller, 0.09 [m] with standard deviation 0.10 [m] for the FL controller, and 0.13 [m] with standard deviation 0.10 [m] for the OC controller. The mean lateral errors are: -0.01 [m] with standard deviation 0.03 [m] for the LMI-based controller, -0.01 [m] with standard deviation 0.08 [m] for the FL controller, and -0.02 [m] with standard deviation 0.06 [m] for the OC controller. The mean orientation errors are: 0.69 [deg] with standard deviation 3.38 [deg] for the LMI-based controller, 1.80 [deg] with standard deviation 4.84 [deg] for the FL controller, and 2.27 [deg] with standard deviation 4.73 [deg] for the OC controller.

Figure 4.7 shows the reference velocities (denoted as v_r^{rf} and v_l^{rf}), the control inputs (referred to as v_r and v_l), and the real track velocities (labeled as "Right Enc." and "Left Enc.") for the case of the LMI-based controller. Two interesting conclusions can be obtained from these plots. First, the motion controller increases the control inputs to compensate the negative slip effect. Second, the low-level PI controllers, responsible to achieve the set-points given by the motion controllers, work properly for both tracks.

Finally, Figure 4.8 displays the pixel displacement values between consecutive images for the case of the LMI-based controller (visual odometry), and the computation time employed (CPU time employed at each sampling instant) by the LMI-based controller and the FL controller. It is important to remark that since the robot is moving on a circular trajectory, different lighting conditions occurred during the motion. This situation leads to shadows in the images employed by the visual odometry approach, which are used to estimate the actual robot velocity. However, as checked in Figure 4.8a, an admissible number of outliers takes place, due to the proper downward camera position and the threshold filter. It is interesting to point out that the mean value of the pixel displacement is around -100 [pixel]. Recall from the experiments presented in Chapter 3, where the robot moved between 0.4 [m/s] and 0.5 [m/s], a pixel displacement of -80 [pixel] was obtained (see Figure 3.11b). Figure 4.8b shows the low computation time employed by the two suggested controllers. Furthermore, it is checked that sampling period is always ensured. In this case, the mean computation time for the LMI-based controller is 0.25 [s] and for the FL controller is 0.20 [s]. Recall that motion controllers run on the computer on-board the mobile robot (Intel Pentium III 1GHz, 512 MB RAM).

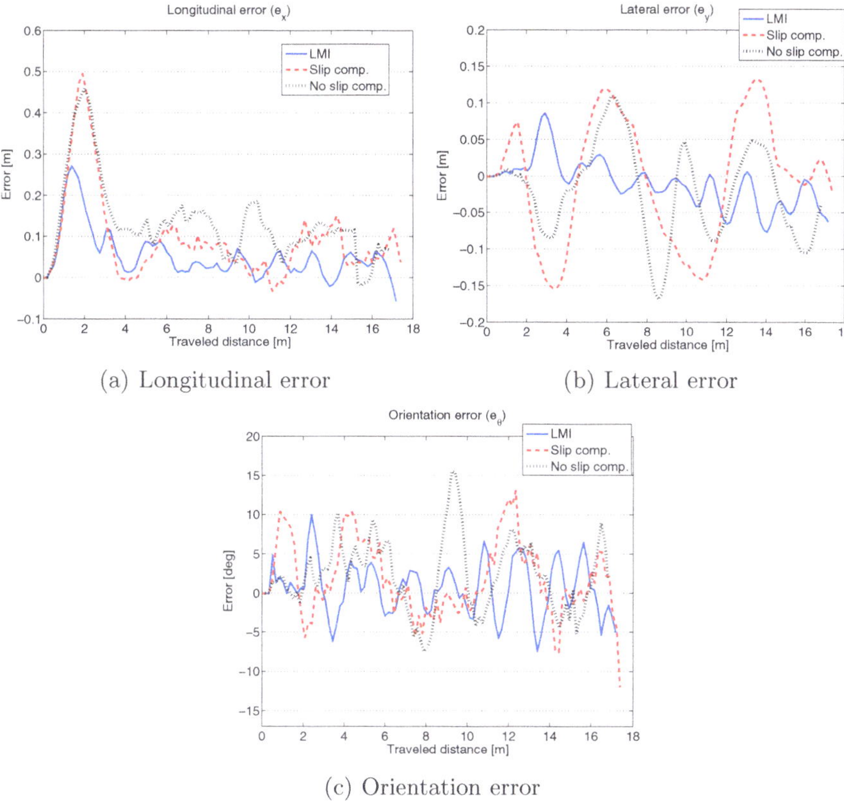

(a) Longitudinal error

(b) Lateral error

(c) Orientation error

Fig. 4.6 Experiment 1. Error with respect to the reference trajectory along the travelled distance

Experiment 2. U-shaped Trajectory

In this second experiment, a U-shaped trajectory has been selected. It constitutes an interesting trajectory which combines straight-line motions and two 90-degrees turns. The total travelled distance was close to 30 [m]. In order to check the navigation architecture at low velocities, a maximum reference velocity of 0.3 [m/s] was selected.

Figure 4.9a shows the reference trajectory and the followed trajectories using the three compared control strategies. Figure 4.9b shows the reference trajectory, the trajectory followed using the LMI-based controller, and the ground-truth (DGPS). In the former plot, it is possible to observe that the trajectory obtained using the LMI-based controller follows properly the reference. The trajectories obtained using the linear feedback controllers have a smooth oscillatory behaviour, especially after the 90-degrees turns. As in previous experiment, the trajectory followed using the LMI-based controller does not fix the ground-truth. Particularly, the maximum lateral deviation is

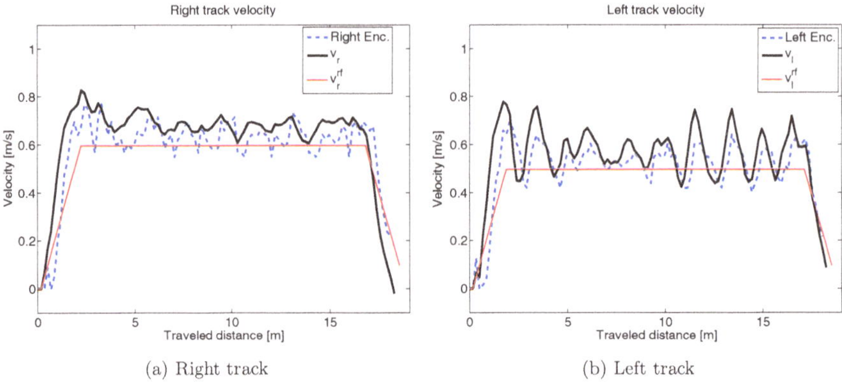

(a) Right track (b) Left track

Fig. 4.7 Experiment 1. Control signals and actual tracks velocities for the LMI-based controller

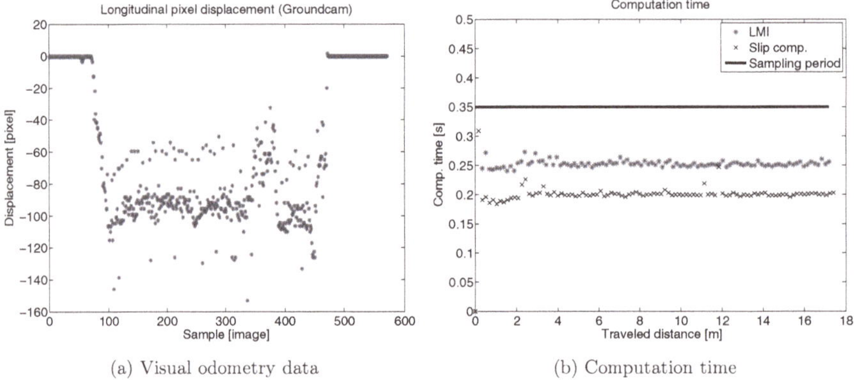

(a) Visual odometry data (b) Computation time

Fig. 4.8 Experiment 1. Pixel displacements (LMI-based controller) and computation times (LMI-based and FL controllers)

0.81 [m], and the mean lateral deviation is 0.28 [m] which means 0.96 [%] of the total travelled distance.

In Figure 4.10a, the orientations are shown. From this figure, it is clearly observed the smooth oscillatory behaviour of the linear feedback controllers, especially at the end of the experiment. The orientation obtained using the LMI-based controller follows the reference with small oscillations. Figure 4.10b plots the reference orientation, the orientation obtained using the LMI-based controller, and the ground-truth (magnetic compass). As expected, the orientation obtained using the LMI-based controller follows the ground-truth with small deviations. In this case, the mean deviation is -1.35 [deg] with standard deviation 5.48 [deg].

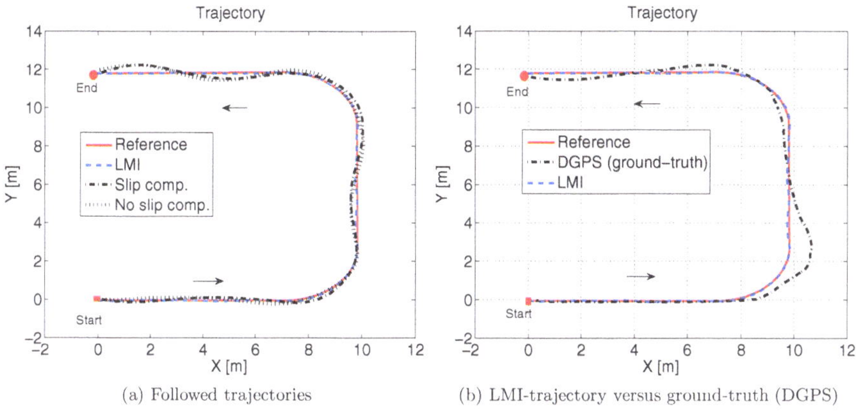

(a) Followed trajectories

(b) LMI-trajectory versus ground-truth (DGPS)

Fig. 4.9 Experiment 2. Followed trajectories during the experiment

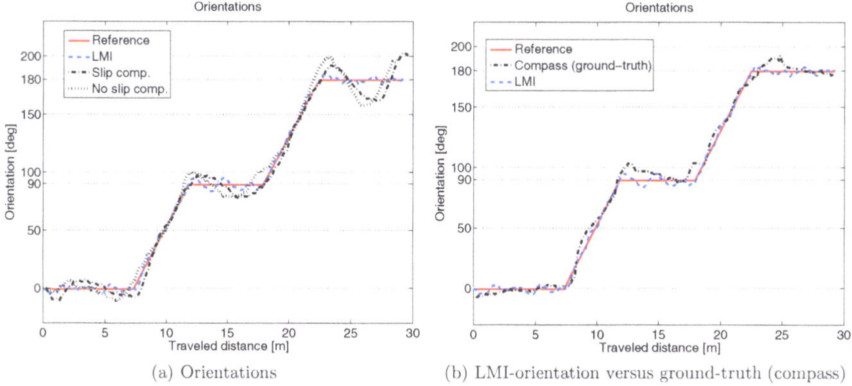

(a) Orientations

(b) LMI-orientation versus ground-truth (compass)

Fig. 4.10 Experiment 1. Orientations

Figure 4.11 displays the slip estimations. In this case, the median slip value for the right track was 7.51 [%], and for the left track was 2.45 [%]. It is possible to notice that the slip for the right track is almost constant. However, for the case of the left track many peaks are observed. Furthermore, it has a mean value different from the right track. This phenomenon was not expected. The explanation for this issue is the smooth oscillatory behaviour of the velocity for the left track, such as it will be explained in the subsequent figures.

The error between the reference trajectory and those steered by the compared controllers are plotted in Figure 4.12. Especially remarkable is the almost zero lateral error achieved by the LMI-based controller. Notice that a substantial oscillatory behaviour is achieved by the linear feedback controllers. Particularly, the OC controller obtains a maximum lateral error close to 0.40 [m]. This error can lead to crash the mobile robot with obstacles in

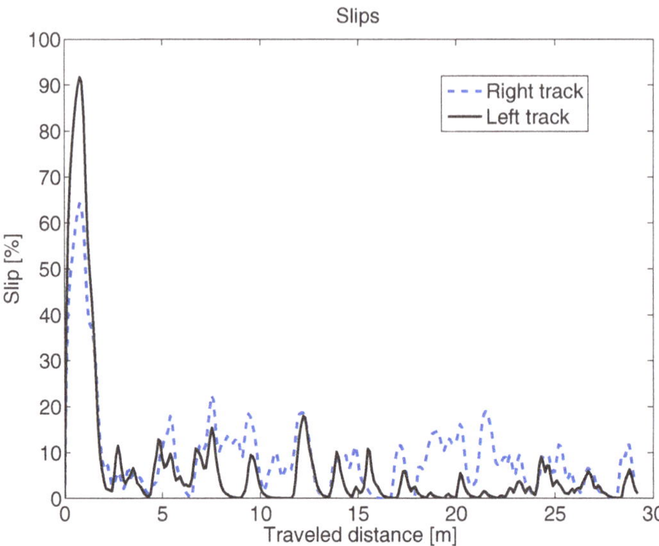

Fig. 4.11 Experiment 2. Slip estimations for the LMI-based controller (comparing the actual robot velocity obtained using visual odometry and track velocities obtained using encoders)

a narrow workspace. The mean longitudinal errors are: 0.02 [m] with standard deviation 0.03 [m] for the LMI-based controller, 0.09 [m] with standard deviation 0.05 [m] for the FL controller, and 0.13 [m] with standard deviation 0.06 [m] for the OC controller. The mean lateral errors are: -0.004 [m] with standard deviation 0.03 [m] for the LMI-based controller, 0.02 [m] with standard deviation 0.17 [m] for the FL controller, and 0.01 [m] with standard deviation 0.18 [m] for the OC controller. The mean orientation errors are: 0.26 [deg] with standard deviation 2.45 [deg] for the LMI-based controller, 0.66 [deg] with standard deviation 8.16 [deg] for the FL controller, and 1.09 [deg] with standard deviation 8.51 [deg] for the OC controller.

Figure 4.13 presents the reference velocities, the control inputs, and the real track velocities for the case of the LMI-based controller. From the control point of view, these velocity profiles are more interesting that those from the previous experiment. In this case, the reference velocity of the left track changes along the trajectory. Again, the right tuning of the PI controllers is noticed, since the real velocities of the tracks follow satisfactorily the set-points. It is important to clarify that the smooth oscillatory behaviour of the left track is stressed by the fact that, such as remarked at the beginning of this subsection, the mobile robot was moving on a partially bumpy gravel soil. In this scenario, little stones produce vibrations to the vehicle, and hence to the tracks. Nevertheless, these small oscillations were expected taking into

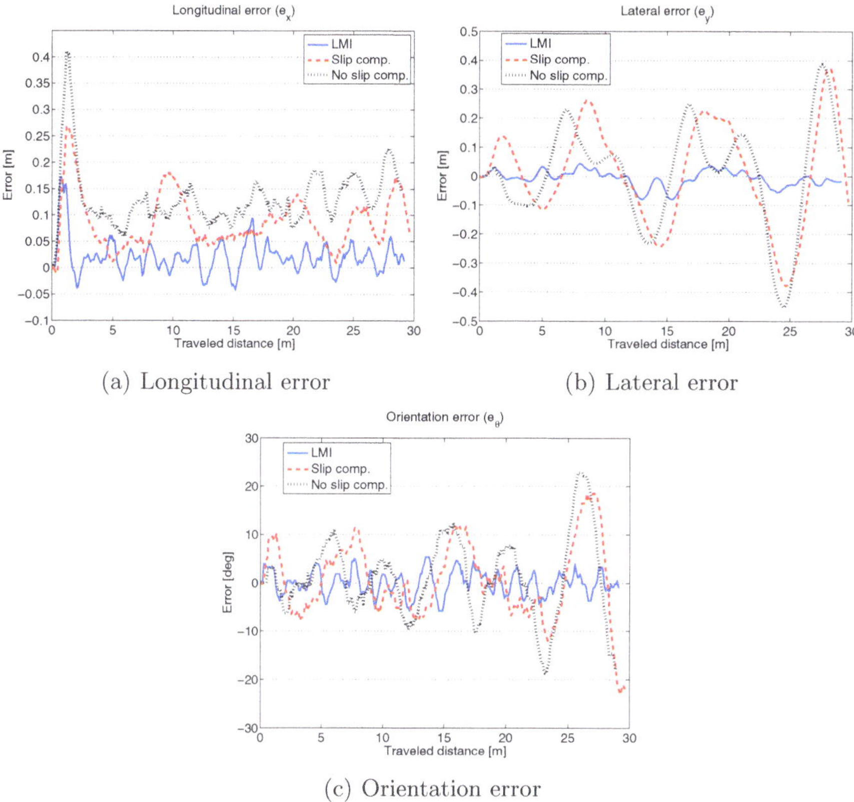

(a) Longitudinal error

(b) Lateral error

(c) Orientation error

Fig. 4.12 Experiment 2. Error with respect to the reference trajectory along the travelled distance

account that a TMR with a limited suspension mechanism has been used as testbed.

Finally, Figure 4.14 displays the pixel displacement values between consecutive images (visual odometry) for the LMI-based controller, and the computation time employed by the LMI-based and the FL controllers. As in the previous experiment, few outliers appear due to the proper downward camera position and the threshold filter. Notice that the mean value of the pixel displacement is around -50 [pixel]. It confirms that the mobile robot was moving at a slower velocity than in previous experiment. Recall that in the previous case, the mean value of the pixel displacement was around -100 [pixel]. Figure 4.14b shows the low computation time employed by the two suggested controllers and the fulfillment of the sampling period. The mean computation time for the LMI-based controller is 0.24 [s] and for the FL controller is 0.19 [s].

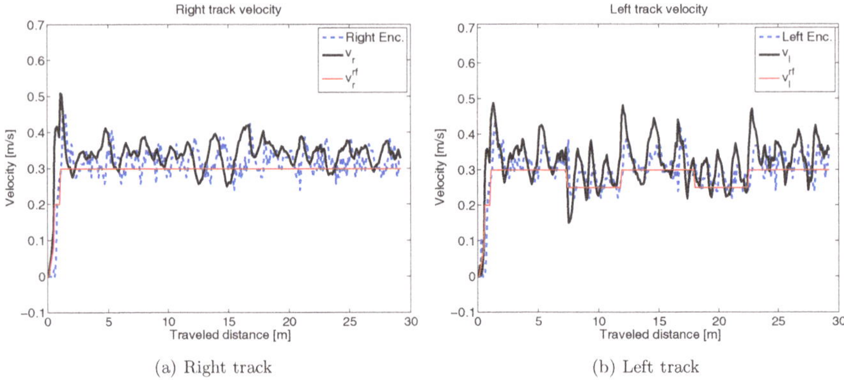

(a) Right track (b) Left track

Fig. 4.13 Experiment 2. Control signals and actual tracks velocities for the LMI-based controller

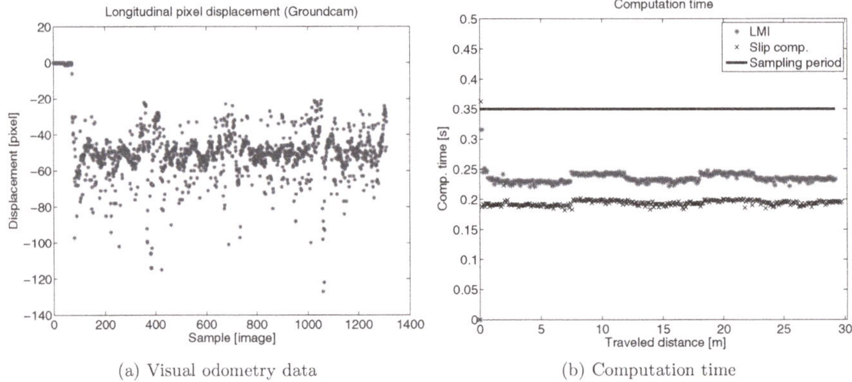

(a) Visual odometry data (b) Computation time

Fig. 4.14 Experiment 2. Pixel displacements (LMI-based controller) and computation time (LMI-based and FL controllers)

4.5 Conclusions

In this chapter, two new control algorithms following diverse paradigms and dealing with the slip phenomenon problem are presented. These controllers are compared through physical experiments. On the one hand, the well-known linear feedback controller [21, 64] has been extended considering slip. For that purpose, time-dependent gains are updated online using the estimated slip. It constitutes an easy understandable and configurable control law that demands a really small computation load.

On the other hand, a more elaborate slip compensation adaptive control scheme using LMI has been designed. In this case, asymptotic stability, performance, and input and state constraints fulfillment are ensured. Furthermore, efficient real-time execution is achieved. For that purpose, a set of feedback gains is obtained offline, one gain for each extreme realization of the system dynamics. Afterwards, an adaptive feedback control law is obtained online as a convex combination of the previously determined gains.

The reported results confirm that the slip compensation motion control laws are suitable for off-road conditions and improve the performance of similar schemes with no slip compensation. Furthermore, the use of more elaborate control strategies, such as the LMI-based adaptive controller, achieves the most satisfactory results in physical experiments.

It is important to remark that solvers of LP problems require relatively small computation resources and execution time. In this case, the LP problem, involving 16 optimization variables, is solved at computation times lower than 0.2 [s]. For that reason, the current online computational aspect can be successfully applied to many other fast systems. Nevertheless, a new selection process to obtain the feedback control gain using geometric relations will be analyzed in future works. For instance, if current parameter realization γ is close to an extreme realization, the offline gain for that extreme realization would be employed at the current sampling instant.

Chapter 5
Robust Predictive Motion Controller for Tracked Robots

This chapter focuses on the design of a robust tube-based predictive controller ensuring robustness (uncertainty in robot position and slip estimation), constraints fulfillment (state and input constraints), stability, and efficient real-time execution. Physical experiments show the better performance of this approach to known solutions.

5.1 Introduction

As commented in previous chapters, when a mobile robot moves in off-road conditions some undesirable effects, such as noisy measurements and inaccurate robot location, are intensified. Additionally, simplified models are usually used for control design purposes. Thus, these factors can lead to uncertainties in the robot motion that may entail inappropriate control actions or even system instability. For that reason, research efforts for the application of robust control strategies in mobile robotics are required. For instance, the work [38] presents a predictive strategy that permits to avoid unexpected static obstacles in the robot workspace. For purposes of MPC real-time execution, a neural network was trained. A Smith-predictor-based generalized predictive controller was discussed in [103]. This formulation permits controlling a mobile robot with dead-time uncertainties taking into account Smith-predictor features in the implementation of the predictive control law. In [25, 26], the trajectory tracking problem for a WMR considering uncertainties in the dynamic model is addressed. The proposed solution is based on sliding mode control. The tracking control problem of a WMR with both parametric and non-parametric uncertainties is also considered in [28]. Here, the controller is designed using the adaptive backstepping technique and neural networks. In [134], a distributed robust control scheme is proposed for formation stabilization of non-holonomic wheeled robots with parametric uncertainty and actuator saturations. The control strategy is based on H_∞ method and is formulated using LMI. The work [127] proposes an adaptive robust fuzzy

R. González, F. Rodríguez, and J.L. Guzmán, *Autonomous Tracked Robots in Planar Off-Road Conditions,* Studies in Systems, Decision and Control 6,
DOI: 10.1007/978-3-319-06038-5_5, © Springer International Publishing Switzerland 2014

control scheme with a genetic algorithm to solve the path tracking problem of a WMR. In this case, uncertainties are included in the dynamic model of the robot.

In order to contribute in this research area, this chapter describes how the adaptive control problem presented in Chapter 4 is formulated as a robust control algorithm (where uncertainties are considered). For this purpose, a robust MPC strategy is considered. This decision is motivated by the fact that MPC constitutes a popular and settled control technique to deal with constrained systems and process uncertainties [8, 18, 96, 99]. From the theoretical point of view, this chapter has addressed the extension of tube-based predictive control to time-varying systems. From a practical point of view, the designed control strategy compensates slip and guarantees that state and input constraints are fulfilled, since these constraints are taken into account in the solution of the optimization problem. Furthermore, an efficient real-time execution is achieved in physical experiments.

An efficient technique for practical implementation of robust MPC is the tube-based MPC. A pioneering work, in which the concept was probably first tackled is [10]. In this work, the design of a robust control law ensuring hard constraints satisfaction is addressed by means of the computation of a sequence of state space regions, called *reachability tube*. The term tube-based refers to those control techniques whose objective is to maintain all the possible trajectories of an uncertain system inside a sequence of admissible regions by using set-theory related tools. Such approach has been widely employed to robustify MPC [24, 75, 94, 95, 126].

Tube-based MPC approaches are motivated by the fact that the predicted evolution of a system obtained using a nominal model differs from the real evolution due to uncertainty. In order to consider this mismatch in the controller synthesis, the basis of tube-based MPC formulation consists in computing the region around the nominal prediction that contains the state of the system under any possible uncertainties [84]. This region can be obtained in two different ways. One possibility is to calculate the region at each step within the prediction horizon. This leads to a sequence of regions, $\{\mathcal{R}_i\}$, called reachable sets, that is, the smallest sets of states of the closed-loop uncertain system that ensure to contain the system state evolution at any time i for any trajectory starting at the origin [24, 41]. The second possibility is to determine a single region that bounds the sequence of reachable sets, that is, $\mathcal{R}_i \subset \mathcal{R}_{i+1} \subset \mathcal{R}$, where usually $\mathcal{R}$ is a robust invariant set [75, 94].

Tube-based MPC usually employs a pre-stabilizing control policy ($u = K(e - \bar{e}) + g$) [24, 45, 94], where the local feedback gain, K, compensates the mismatch between the nominal, $\bar{e}$, and the real, e, evolution of the system [24, 75, 84], and a deterministic MPC controller, g, is employed to stabilize the nominal system for which tighter constraint sets are considered. This chapter details one way to implement a robust tube-based MPC controller by means of reachable sets. For experimental validation a tracked mobile robot has been used.

This chapter is organized as follows. Section 5.2 concerns with problem statement and control objectives. In Section 5.3, the robust tube-based MPC strategy is described. Section 5.4 demonstrates the performance of the proposed control approach through physical experiments. Finally, Section 5.5 presents some conclusions and future research.

5.2 Problem Statement

This chapter deals with the discrete-time trajectory tracking error model with additive uncertainty (2.28) and the nominal trajectory tracking error model (2.30). The mismatch between the nominal and the real evolution of the system is given by

$$\tilde{e}(k) = e(k) - \bar{e}(k), \tag{5.1}$$

where $\tilde{e} = [\tilde{e}_x \quad \tilde{e}_y \quad \tilde{e}_\theta]^T \in \mathbb{R}^3$ is the mismatch state.

Then, the control objective is to compensate this mismatch and to steer the nominal system as close as possible to the real process without constraints violation. For that purpose, the following pre-stabilizing control policy is considered [24, 45]

$$u(k) = K\tilde{e}(k) + g(k), \tag{5.2}$$

where K is a local controller whose goal is to compensate the mismatch. Recall from Chapter 2 that, $u = [u_1 \quad u_2]^T$ is the input vector of (2.28), and $g = [g_1 \quad g_2]^T$ is the control input for the nominal system (2.30).

Finally, the dynamics of the closed-loop mismatch system is defined replacing (2.28) and (2.30) into (5.1), what results in

$$\tilde{e}(k+1) = e(k+1) - \bar{e}(k+1) = \big(A_v(k) + B_d K\big)\tilde{e}(k) + w(k), \tag{5.3}$$

where $A_v(k)$ was given by (2.29) and B_d was defined in (2.27).

The main features of the suggested robust tube-based MPC strategy are:

- Robustness. Following the tube-based MPC policy, additive uncertainties and time-varying dynamics are taken into account in the control design.
- Performance. An optimization problem (QP) is solved at each sampling instant obtaining the proper control actions as a compromise between small deviations from the reference trajectory and suitable control actions.
- Input and state constraints fulfillment. This requirement is guaranteed by ensuring constraints satisfaction in the minimization of the MPC control law.

- Stability[1]. It is ensured through a quadratic Lyapunov function constituting the terminal cost for the nominal system (nominal MPC) and a robust positively invariant set.
- Efficient real-time execution. A standard nominal MPC is solved online to control the nominal system. This fact implies that tube-based MPC strategies fit properly to applications with fast dynamics and where high sampling frequencies are employed.

5.3 Robust Tube-Based MPC Controller

This section describes the implementation of a robust tube-based MPC controller considering the ideas proposed in [24, 94]. First, the local control law, K, that compensates the effect of the mismatch (5.1) has to be calculated. Afterwards, reachable sets containing the real state for every possible realization of the uncertainty and every admissible parameters occurrence are calculated. Finally, online computation is devoted to the solution of a standard MPC for the nominal system with restricted constraints.

Figure 5.1 shows the robust tube-based MPC strategy using offline reachable sets. First, the reachable sets for the local uncertain closed-loop system (5.3) are calculated. Then the state and input constraints are replaced, using the previously determined reachable sets. Online computation is only devoted to a standard MPC controller handling the nominal system for which tighter constraints are considered. Finally, LMI are employed to determine the terminal cost for the MPC optimization problem and a terminal robust invariant set is calculated (see Subsection 5.3.3).

The steps that will be explained subsequently to implement the robust tube-based MPC approach are:

1. Local control gain (Subsection 5.3.1). The local control law, K, that compensates the effect of the mismatch (5.1) has to be calculated. In particular, such local gain tries to compensate the system realization A_υ and it will have the form

   $$\tilde{K} = K^n + K^r \Delta v_r^{rf} + K^l \Delta v_l^{rf},$$

   where each component K^n, K^r and K^l will try to compensate a part of the system realization.

2. Reachable sets (Subsection 5.3.2). Reachable sets containing the real state for every possible realization of uncertainty are calculated. In this case, such reachable sets will also include the bounding set dealing with the

[1] Notice that, it is only ensured the asymptotic stability for the nominal MPC related to the nominal system. It is considered that the problem of ensuring stability of the whole tube-based control strategy is beyond the scope of this monograph.

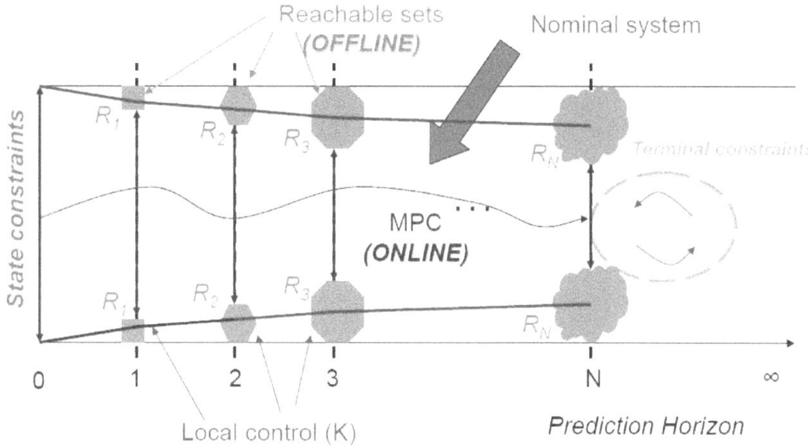

Fig. 5.1 Robust tube-based MPC control strategy. Reachable sets are solved offline compensating uncertainty, online computation is devoted to solve a nominal MPC

system variation which has not been compensated by the local control gain. In this way, reachable sets will have the form

$$\mathcal{R}_{k+i+1} \triangleq \bigcup_{A_v \in \mathcal{A}^v} (A_v + B_d\tilde{K})\mathcal{R}_{k+i} \oplus W, \quad \forall i = 0, \ldots, N-1, \quad (5.4)$$

where $\mathcal{R}_{k+i}$ is the reachable set of the i-th step within the prediction horizon N.

3. Modification of state and input constraints (Subsection 5.3.2). In order to take into account the "control effort" spent to compensate the system mismatch and the uncertainty, original state and input constraints are replaced with more restricted ones using the previously determined reachable sets

$$\bar{E}_i = E \ominus \mathcal{R}_i \quad \forall i = 1, \ldots, N, \quad (5.5)$$

for the state constraints, and for the case of the input constraints

$$\bar{U}_i = U \ominus \left(K^n\mathcal{R}_i \oplus K^r \Delta v_r^{ref}\mathcal{R}_i \oplus K^l \Delta v_l^{ref}\mathcal{R}_i\right) \forall i = 0, \ldots, N-1. \quad (5.6)$$

4. Terminal constraints for MPC (Subsection 5.3.3).
5. MPC strategy for the nominal system (Subsection 5.3.4).

5.3.1 Local Compensation of System Dynamics

This subsection describes how the control law for the local uncertain system (5.3) is determined. In this case, the local closed-loop system without additive uncertainty ($W = \{0\}$) is considered, since this term will be included in the reachable sets (see Subsection 5.3.2).

In order to obtain a local control law for the closed-loop system (5.3) without the additive uncertainty term, first, the system matrix A_υ is decomposed into a time-invariant part and a time-varying one (the dependence on time k has been omitted)

$$A_\upsilon = I_3 + A^r v_r^{rf} + A^l v_l^{rf}, \tag{5.7}$$

where

$$A^r = \begin{bmatrix} 0 & \epsilon_r & 0 \\ -\epsilon_r & 0 & \chi \\ 0 & 0 & 0 \end{bmatrix}, \quad A^l = \begin{bmatrix} 0 & -\epsilon_l & 0 \\ \epsilon_l & 0 & \chi \\ 0 & 0 & 0 \end{bmatrix}, \tag{5.8}$$

with $\epsilon_r = T_s\left(\frac{1-\bar{i}_r}{b}\right)$, $\chi = \frac{T_s}{2}$ and $\epsilon_l = T_s\left(\frac{1-\bar{i}_l}{b}\right)$.

Now, from Assumption 1 (see Chapter 2), it is obtained that

$$v_r^{rf} = \bar{v}_r^{rf} + \Delta v_r^{rf} \quad \Rightarrow \quad \Delta v_r^{rf} \in [\delta v_r^{rf,m}, \delta v_r^{rf,M}], \tag{5.9}$$

$$v_l^{rf} = \bar{v}_l^{rf} + \Delta v_l^{rf} \quad \Rightarrow \quad \Delta v_l^{rf} \in [\delta v_l^{rf,m}, \delta v_l^{rf,M}], \tag{5.10}$$

where $\bar{v}_r^{rf}$ and $\bar{v}_l^{rf}$ are the nominal reference velocities of the right and left tracks, respectively, and Δv_r^{rf} and Δv_l^{rf} are a new range in the reference velocities. Then, the system matrix A_υ is expressed as

$$A_\upsilon = A^n + A^r \Delta v_r^{rf} + A^l \Delta v_l^{rf}, \tag{5.11}$$

where $A^n = I_3 + A^r \bar{v}_r^{rf} + A^l \bar{v}_l^{rf}$ represents the time-invariant part of A_υ and the rest of terms are time-varying that must be bounded. Replacing (5.11) into (5.3), it follows

$$\tilde{e}(k+1) = \left(A^n + A^r \Delta v_r^{rf} + A^l \Delta v_l^{rf} + B_d K\right)\tilde{e}(k) + w(k). \tag{5.12}$$

In order to compensate the system dynamics, the local control gain can be defined as

$$\tilde{K} = K^n + K^r \Delta v_r^{rf} + K^l \Delta v_l^{rf}, \tag{5.13}$$

which implies that equation (5.12) becomes

$$\tilde{e}(k+1) = A_{cl}^n \tilde{e}(k) + A_{cl}^r \Delta v_r^{rf} \tilde{e}(k) + A_{cl}^l \Delta v_l^{rf} \tilde{e}(k) + w(k), \tag{5.14}$$

where $A_{cl}^n = (A^n + B_d K^n)$, $A_{cl}^r = (A^r + B_d K^r)$, and $A_{cl}^l = (A^l + B_d K^l)$. The gain K^n is directly obtained solving an LQR problem. The gains K^r and K^l have to be chosen such that the effect of the system dynamics in matrices A^r and A^l is partially compensated in closed-loop. In this way

$$K^r = \begin{bmatrix} 0 & -\frac{\epsilon_r}{T_s} & 0 \\ 0 & 0 & 0 \end{bmatrix}, \quad K^l = \begin{bmatrix} 0 & \frac{\epsilon_l}{T_s} & 0 \\ 0 & 0 & 0 \end{bmatrix}, \tag{5.15}$$

and, it holds

$$A_{cl}^r = \begin{bmatrix} 0 & 0 & 0 \\ -\epsilon_r & 0 & \chi \\ 0 & 0 & 0 \end{bmatrix}, \quad A_{cl}^l = \begin{bmatrix} 0 & 0 & 0 \\ \epsilon_l & 0 & \chi \\ 0 & 0 & 0 \end{bmatrix}. \tag{5.16}$$

Note that it is possible to completely remove the effects of the system dynamics in the $\tilde{e}_x$- and $\tilde{e}_\theta$-components of the state. However, the $\tilde{e}_y$-component cannot be compensated directly. For that reason, the bounds of the terms ϵ_r, ϵ_l and χ have to be considered in the offline reachable sets calculation. This issue is addressed in Subsection 5.3.2.

5.3.2 Reachable Sets Calculation

This subsection focuses on the reachable sets calculation.

Consider the uncertain closed-loop system (5.3) whose evolution depends on the local control action $(\tilde{K}\tilde{e})$ and the uncertainties. Then, the reachable set at first step within the prediction horizon is denoted as $\mathcal{R}_k = \{0\}$, where subscript k means current sampling instant (first step within the prediction horizon). The rest of reachable sets are recursively calculated as

$$\mathcal{R}_{k+i+1} \triangleq \bigcup_{A_\upsilon \in \mathcal{A}^\upsilon} (A_\upsilon + B_d \tilde{K})\mathcal{R}_{k+i} \oplus W, \quad \forall i = 0, \dots, N-1, \tag{5.17}$$

where $\mathcal{R}_{k+i}$ is the reachable set of the i-th step within the prediction horizon N (see Assumption 4). In this particular case, the reachable sets are computed as

$$\begin{aligned} \mathcal{R}_{k+i+1} = A_{cl}^n \mathcal{R}_{k+i} \oplus co\{(A_{cl}^r \delta v_r^{rf,m} \mathcal{R}_{k+i}) \cup (A_{cl}^r \delta v_r^{rf,M} \mathcal{R}_{k+i})\} \oplus \\ \oplus co\{(A_{cl}^l \delta v_l^{rf,m} \mathcal{R}_{k+i}) \cup (A_{cl}^l \delta v_l^{rf,M} \mathcal{R}_{k+i})\} \oplus W, \\ \forall j = 0, \dots, N-1, \end{aligned} \tag{5.18}$$

where $co\{\cdot\}$ is the convex hull.

Notice that, in order to determine the offline reachable sets, bounds in the elements of the matrices A_{cl}^r and A_{cl}^l have been considered. Recall from (5.14)-(5.16) that these elements cannot be fully compensated by the local control law.

Finally, following the ideas from [24], [75], [83], [94], original constraints (2.34) are replaced with more restricted ones using the offline reachable sets previously calculated. That is,

$$\bar{E}_i = E \ominus \mathcal{R}_i \quad \forall i = 1, \ldots, N. \tag{5.19}$$

Then, the state constraints satisfaction is ensured and feasibility is also preserved in the presence of uncertainties in the system (2.28). Additionally, the input constraints are also replaced by

$$\bar{U}_i = U \ominus \left(K^n \mathcal{R}_i \oplus K^r \Delta v_r^{ref} \mathcal{R}_i \oplus K^l \Delta v_l^{ref} \mathcal{R}_i\right) \quad \forall i = 0, \ldots, N-1 \tag{5.20}$$

Assumption 4. *Assume that $\mathcal{R}_\infty \subset E$ and $\mathcal{R}_\infty \subset E$. It ensures the sets $\bar{E}_i$ and $\bar{U}_i$ are not empty.*

5.3.3 Terminal Constraints for MPC

A common approach to ensure the stability of deterministic MPC consists in incorporating both a terminal cost (Φ) and a terminal constraint set (Ω) [96]. In this subsection, the local control law that supposes the terminal cost for the nominal system and the terminal robust invariant set are detailed. Notice that, similar stability properties for analogous control strategies are found in the literature [24, 75, 84, 94].

The purpose of the terminal cost is to ensure closed-loop stability. To this end, it requires the use of a Lyapunov function with a stabilizing control law. In this case, the terminal cost, Φ, is given by the Lyapunov function $V(e) = e^T P e$. This mathematical development goes beyond the scope of this book, but for details one can check [40].

Once the terminal cost is discussed, the second property to ensure the stability of the MPC is that the last element of the predicted state sequence must belong to an invariant set [96]. In this case, a similar approach to that found in [12, 37, 68] is followed to obtain the maximal robust invariant set for the uncertain system contained in the state and input constraint sets. For that purpose, the concept of one-step operator is employed as

$$Q^j_{\mathcal{A}}(\Omega) = \{e \in E : K(A^j_v)e \in U, \quad (A^j_v + B_d K(A^j_v))e + w \in \Omega, \ \forall w \in W\}. \tag{5.21}$$

Note that the one-step operator is a standard tool for the invariant sets calculation through iterative procedures [34], [68], [82].

Then, taking into account the previously defined one-step operator, the maximal robust invariant set for the uncertain system (2.28) is obtained by means of the following iterative procedure:

1. Initialization: $\Omega_0 = E \cap \{\omega \in \mathbb{R}^3 : \ K(A^j_v)\omega \in U, \ \forall j = 1, \ldots, n_v\}$.
2. Iteration: $\Omega_{k+1} = \Omega_k \cap Q^j_{\mathcal{A}}(\Omega_k) \ \forall j = 1, \ldots, n_v$.

3. Termination condition: stop when $\Omega_{k+1} = \Omega_k$ or $\Omega_{k+1} = \emptyset$. Set $\Omega = \Omega_\infty = \Omega_{k+1}$.

An important issue when dealing with an algorithmic procedure for computing the robust invariant sets is its finite determinedness, that is, the conditions under which the algorithm provides a solution after a finite number of iterations. Results regarding the problem of finite determination can be found in [12, 37, 68]. Notice that, finite determinedness has not been proved for this case, since it is beyond the scope of this monograph. Nevertheless, the application of this algorithm to the case under analysis provides a result after a finite number of iterations.

5.3.4 MPC Strategy

Finally, the online computational aspect related to MPC is considered. Notice that MPC policy deals with the nominal system (2.30), since the mismatch between the real system and the nominal one is compensated by the local control law and uncertainty is within the reachable sets. Assume that a measurement of the state $\bar{e}$ is available at the current time k. Then, the optimization problem is stated as follows

$$\min_{G(k)\triangleq\{g_k,\ldots,g_{k+N-1}\}} J_N(\bar{e}(k), G(k)) = \sum_{i=0}^{N-1} \bar{e}_{k+i}^T Q_M \bar{e}_{k+i} + g_{k+i}^T R_M g_{k+i} + \Phi(\bar{e}_{k+N}) \tag{5.22}$$

$$\begin{aligned} \text{subject to} \quad & \bar{e}_{k+i} \in \bar{E}_i && \forall i = 1, \ldots, N, \\ & g_{k+i} \in \bar{U}_i && \forall i = 0, \ldots, N-1, \\ & \bar{e}_{k+N} \in \Omega \ominus \mathcal{R}_N, \end{aligned}$$

where $\bar{e}_{k+i}$ denotes the predicted state vector at time $k + i$, obtained by applying the input sequence $G(k) \triangleq \{g_k, \ldots, g_{k+N-1}\}$ to model (2.30) starting from the state $\bar{e}(k)$. The terminal cost, $\Phi(\cdot)$, and the terminal constraint set, given by the region Ω, were both calculated in previous subsection.

Note that the MPC includes the new state and input constraints, $\bar{E}_i$ and $\bar{U}_i$, come from (5.19)-(5.20). Finally, matrices $Q_M = Q_M^T \geq 0$ and $R_M = R_M^T > 0$, constitute parameters to be tuned for the MPC control law. Depending on their values, more attention can be given to the states or to the control signals (see a discussion about the selected values in Section 5.4).

To summarize, MPC control law is based on the following idea [7]. At time k, compute the optimal solution $G^*(k) = \{g_k^*, \ldots, g_{k+N-1}^*\}$ to problem (5.22), and applying

$$g(k) = g_k^*, \tag{5.23}$$

as input to system (2.30); repeat the optimization (5.22) at time $k+1$ based on the new state $\bar{e}(k+1)$, and continue iteratively.

5.4 Results

This section analyzes the performance of the proposed robust tube-based MPC approach through physical experiments. Furthermore, the slip compensation adaptive controller using LMI and the slip compensation adaptive feedback controller using dynamic feedback linearization, both presented in Chapter 4, have been considered for comparison purposes. Additionally, the original formulation of the linear feedback controller, detailed in [21, 64], has also been implemented. From now on, the robust tube-based MPC with offline reachable sets calculation is denoted as Offline MPC, the slip compensation adaptive controller using LMI is referred to as LMI-based controller, the slip compensation adaptive controller using dynamic feedback linearization is referred to as FL controller, and the original linear feedback controller as OC controller.

The parameters used for the physical experiments are: $b = 0.5[m]$, nominal slip is 0.10 (10 [%]), the uncertainty set is given by $W = \{w_1, w_2 \in \pm 0.0035[m], w_3 \in \pm 0.25[deg]\}$. State constraints are $E = \{e_x, e_y \in \pm 0.5[m], e_\theta \in \pm 20[deg]\}$, reference linear track velocities are restricted to $\{v_r^{rf}, v_l^{rf} \in [0.1, 1.4][m/s]\}$, and real linear track velocities are restricted to $\{v_r, v_l \in [-2, 2][m/s]\}$, $Q_O = Q_Y = diag([1 \quad 1 \quad 0.0001])$ and $R_O = R_Y = I_2$, $Q_M = diag([1 \quad 1 \quad 0.0001])$ and $R_M = I_2$. The parameters of the linear feedback controllers are set to $\beta = 1$ and $\xi = 0.6$ in order to reach a soft overdamped closed-loop behaviour. Notice that similar configuration parameters to that used in previous chapter and in Chapter 2, Subsection 2.7.3 have also been employed here. The sampling period has been selected as $T_s = 0.35[s]$. The values of matrices $Q_M = diag([q_m^1 \quad q_m^2 \quad q_m^3])$ and $R_M = diag([r_m^1 \quad r_m^2])$ have been chosen as $Q_M = diag([1 \quad 1 \quad 0.0001])$ and $R_M = I_2$, in order to obtain an appropriate performance, small errors and smooth control signals. The initial location of the mobile robot is always $[0 \quad 0 \quad 0]^T$.

5.4.1 Experiment 1. Circular Trajectory

Figures 5.2a show the reference trajectory and the followed trajectories using the four control strategies. Figure 5.2b shows the reference trajectory, the trajectory followed using the predictive controller, and the ground-truth (DGPS). First, it is possible to observe that the predictive and LMI-based controllers achieve the best behaviour in comparison to the rest of motion controllers. Nevertheless, all the compared controllers achieves a proper result and the mobile robot comes back to the initial point of the circular trajectory. Notice that again, from Figure 5.2b, the trajectory obtained using the predictive controller does not fix the ground-truth. In this particular experiment, the maximum lateral deviation is 0.53 [m], and the mean lateral deviation is 0.18 [m] which means 1 [%] of the total travelled distance. Recall that this phenomenon is due to the use of the extended kinematic model (2.5) to estimate the robot location (see a deeper discussion at the beginning of Section 4.4 in Chapter 4).

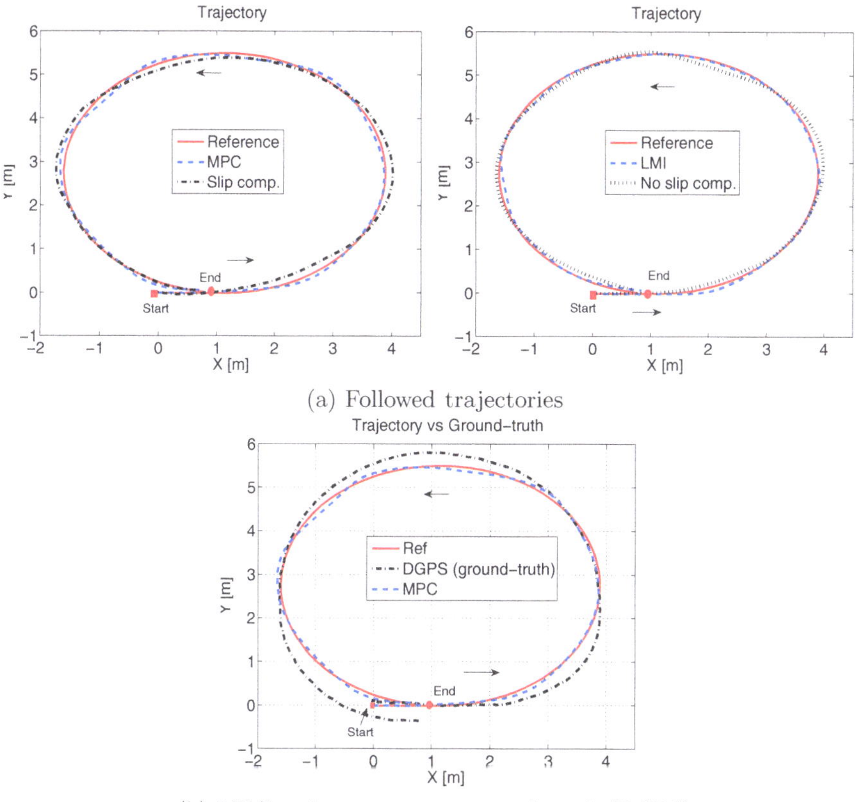

(a) Followed trajectories

(b) MPC-trajectory versus ground-truth (DGPS)

Fig. 5.2 Experiment 1. Followed trajectories during the experiment

Figures 5.3a show the orientations followed by the four compared control approaches. It is observed the satisfactory behaviour of the compared controllers, especially the almost constant behaviour of the predictive controller. Figure 5.3b plots the reference orientation, the orientation obtained using the MPC controller, and the ground-truth (magnetic compass). In this case, the orientation obtained using the MPC controller follows more accurately the ground-truth than in the case of the LMI-based controller (see Figure 4.4b in previous chapter). Particularly, the mean deviation for the MPC controller is -0.67 [deg] with a standard deviation 8.26 [deg].

Figure 5.4 displays the slip estimations. As explained in the previous chapter, the mean values of both slip estimations are considered. The mean slip

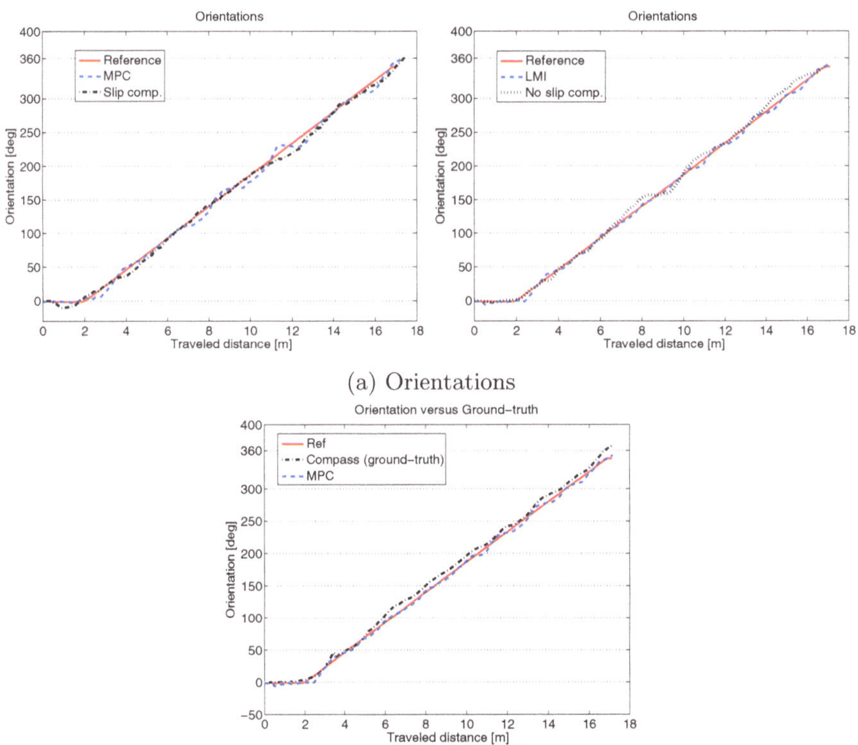

(b) MPC-orientation versus ground-truth (compass)

Fig. 5.3 Experiment 1. Orientations

value is 4.81 [%] that confirms the results presented in Chapter 2 about the mean slip value on a gravel soil.

The errors between the reference trajectory and those steered by the compared controllers are displayed in Figure 5.5. In these plots is verified the proper behaviour of the predictive and the LMI-based controllers. The suggested FL controller also achieves a satisfactory result in comparison to the OC controller. A positive longitudinal error is observed. This means that the real robot is pursuing the reference robot. The mean longitudinal errors are: 0.05 [m] with standard deviation 0.05 [m] for the MPC controller, 0.05 [m] with standard deviation 0.06 [m] for the LMI-based controller, 0.09 [m] with standard deviation 0.10 [m] for the FL controller, and 0.13 [m] with standard deviation 0.1 [m] for the OC controller. The mean lateral errors are: -0.002 [m] with standard deviation 0.05 [m] for the MPC controller, -0.01 [m] with standard deviation 0.03 [m] for the LMI-based controller, -0.01 [m] with standard deviation 0.08 [m] for the FL controller, and -0.02 [m] with standard deviation 0.06 [m] for the OC controller. The mean orientation errors are: 0.50 [deg] with standard deviation 5.80 [deg] for the MPC controller, 0.69

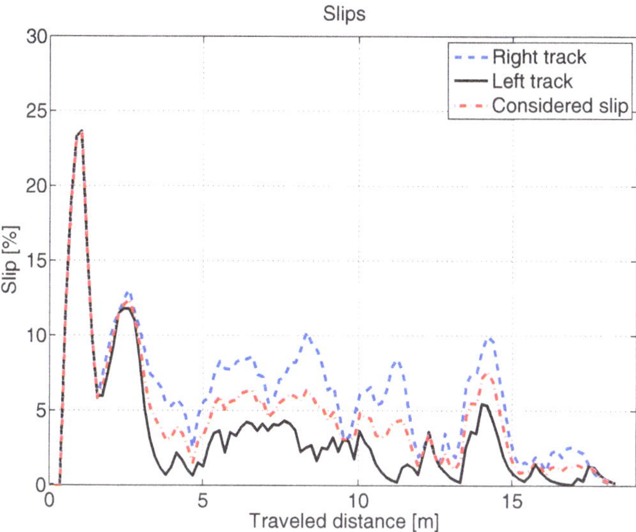

Fig. 5.4 Experiment 1. Slip estimations for the MPC controller (comparing the actual robot velocity obtained using visual odometry and track velocities obtained using encoders)

[deg] with standard deviation 3.38 [deg] for the LMI-based controller, 1.80 [deg] with standard deviation 4.84 [deg] for the FL controller, and 2.27 [deg] with standard deviation 4.73 [deg] for the OC controller. It is interesting to remark the proper results obtained by the most elaborate control strategies, namely, the predictive and the adaptive schemes. Particularly for this experiment, although the predictive approach obtains the smaller mean errors, the LMI-based controller achieves a surprising almost zero lateral error (look at the standard deviation).

Figure 5.6 shows the reference velocities (denoted as v_r^{rf} and v_l^{rf}), the control inputs (referred to as v_r and v_l), and the real track velocities (labeled as "Right Enc." and "Left Enc.") for the case of the MPC controller. The motion controller increases the control inputs to compensate the negative slip effect. Furthermore, in these plots, it can be checked the right tuning of the low-level PI controllers, that is, the real velocities achieved by the tracks follow properly the set-points given by the motion controller.

Finally, Figure 5.7 displays the pixel displacement values between consecutive images for the case of the MPC controller (visual odometry), and the computation time employed by the predictive controller, the LMI-based controller, and the FL controller. From Figure 5.7a it is possible to observe that the pixel displacement values are aligned mainly around -100 [pixel]. It means that for this challenging trajectory in which the robot changes constantly the orientation and many shadows occur, the visual odometry algorithm works satisfactorily. Figure 5.7b shows the computation time employed by the the predictive controller, and the two suggested controllers in previous chapter.

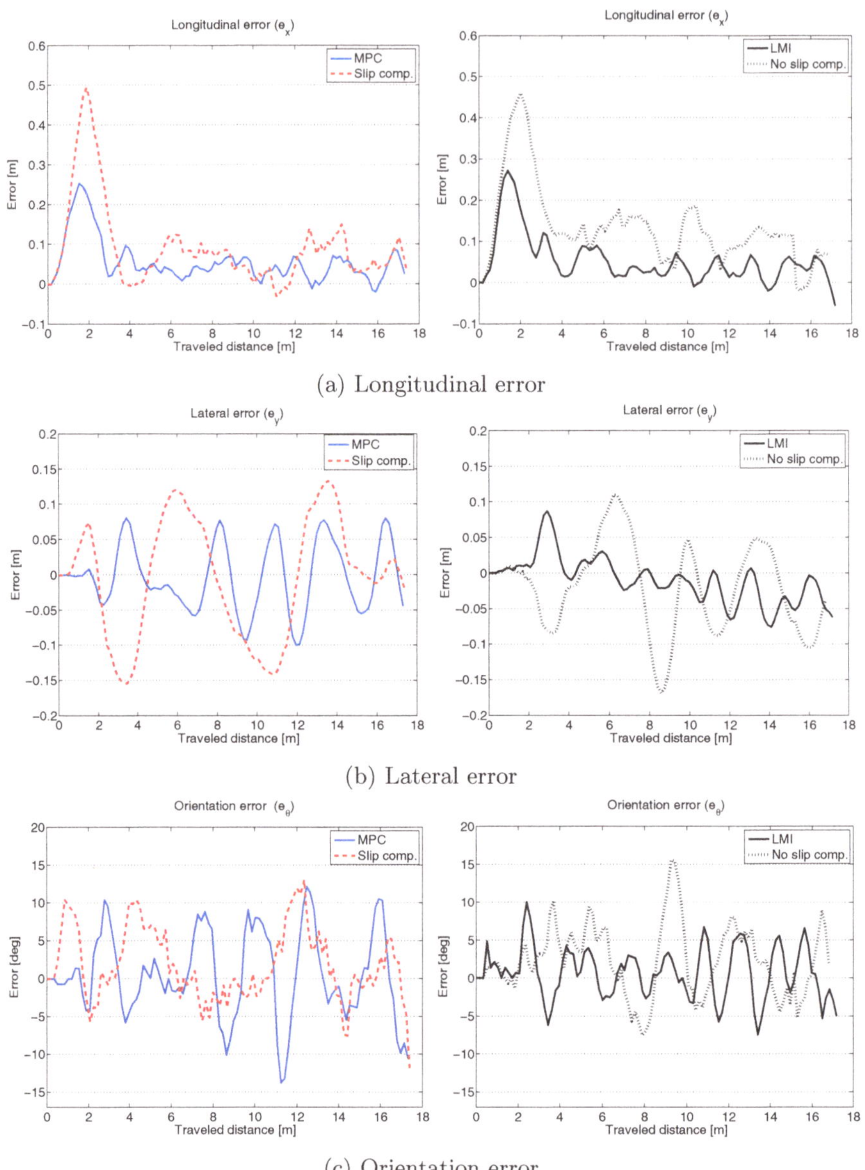

(a) Longitudinal error

(b) Lateral error

(c) Orientation error

Fig. 5.5 Experiment 1. Error along the travelled distance

In this case, the mean computation time for the predictive controller is 0.24 [s], for the LMI-based controller is 0.25, and for the FL controller is 0.20 [s]. Recall that the motion controllers ran on the computer on-board the mobile robot (Intel Pentium III 1GHz, 512 MB RAM).

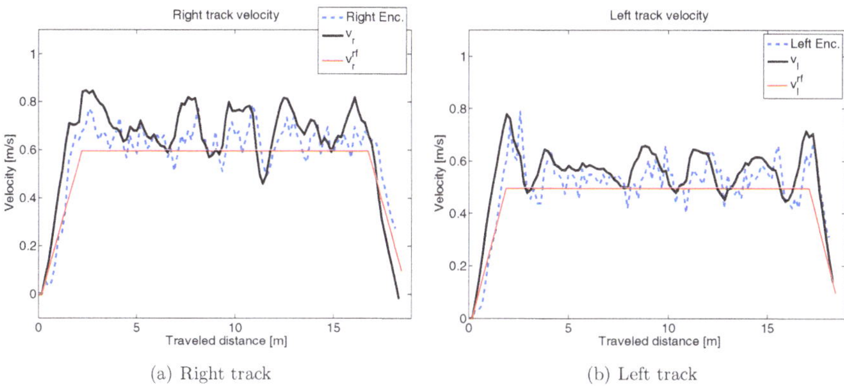

(a) Right track (b) Left track

Fig. 5.6 Experiment 1. Control signals and actual tracks velocities for the MPC controller

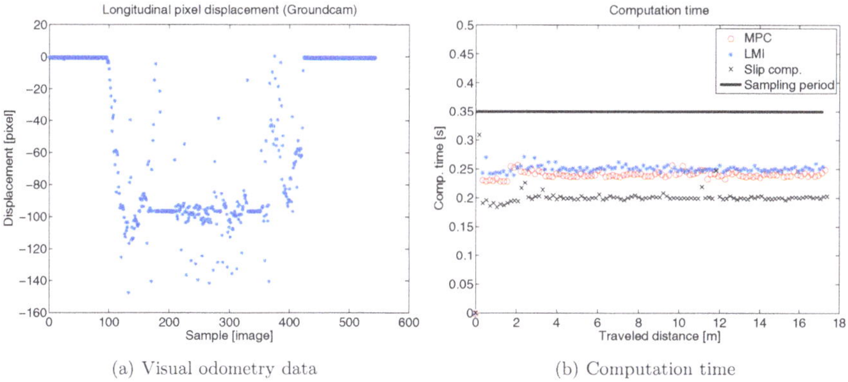

(a) Visual odometry data (b) Computation time

Fig. 5.7 Experiment 1. Pixel displacements (for the MPC controller) and computation time (CPU time employed at each sampling instant)

5.4.2 Experiment 2. U-shaped Trajectory

In this second experiment, a U-shaped trajectory has been selected. It constitutes an interesting trajectory that combines straight-line motions and two 90-degrees turns. The total travelled distance was close to 30 [m]. In order to check the navigation architecture at low velocities, a maximum reference velocity of 0.3 [m/s] was selected.

Figures 5.8a show the reference trajectory and the followed trajectories using the four control strategies. Figure 5.8b displays the reference trajectory, the trajectory followed using the predictive controller, and the ground-truth (DGPS). In the former plot, it is possible to observe that the trajectory obtained using the predictive controller follows satisfactorily the reference. The trajectories obtained using the linear feedback controllers have an smooth oscillatory behaviour, especially after the 90-degrees turns. As in the previous experiment, the trajectory followed using the predictive controller does not fix the ground-truth. Particularly, the maximum lateral deviation is 0.72 [m], and the mean lateral deviation is -0.11 [m], what means 0.40 [%] of the total travelled distance.

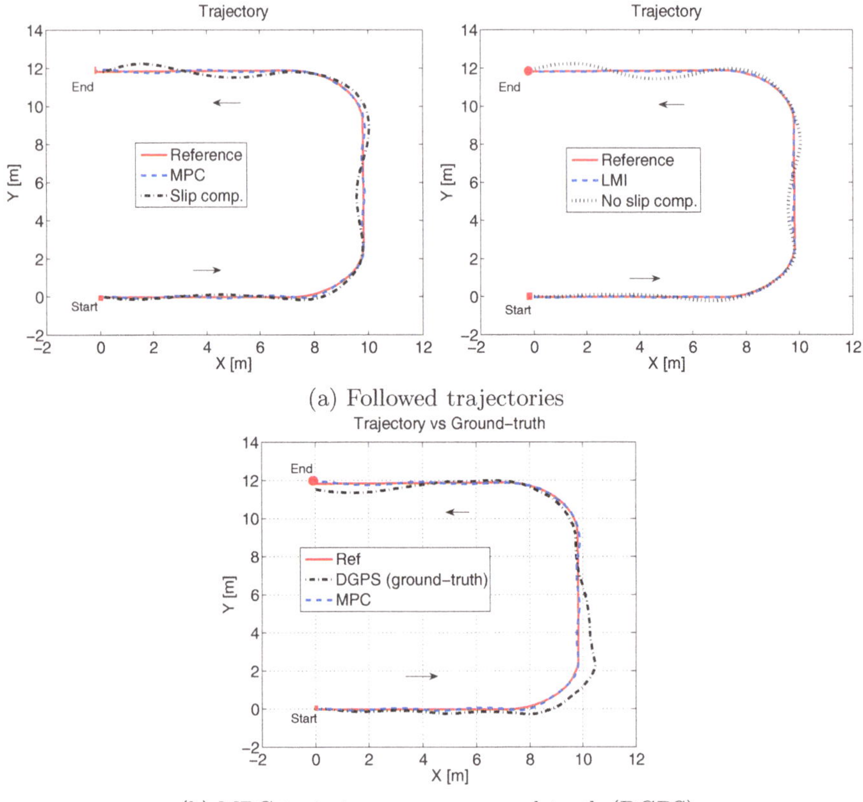

(a) Followed trajectories

(b) MPC-trajectory versus ground-truth (DGPS)

Fig. 5.8 Experiment 2. Followed trajectories during the experiment

Figures 5.9a show the orientations. From these figures, it is clearly observed the smooth oscillatory behaviour of the linear feedback controllers, especially at the end of the experiment. The orientation obtained using the predictive controller follows the reference with very small oscillations. Figure 5.9b plots the reference orientation, the orientation obtained using the predictive controller, and the ground-truth (magnetic compass). As expected, the orientation obtained using the predictive controller does not fix the ground-truth. In this case, the mean deviation is -0.11 [deg] with standard deviation 6.47 [deg].

Figure 5.10 displays the slip estimations. In this case, the median slip value for the right track was 6.32 [%], and for the left track was 2.17 [%].

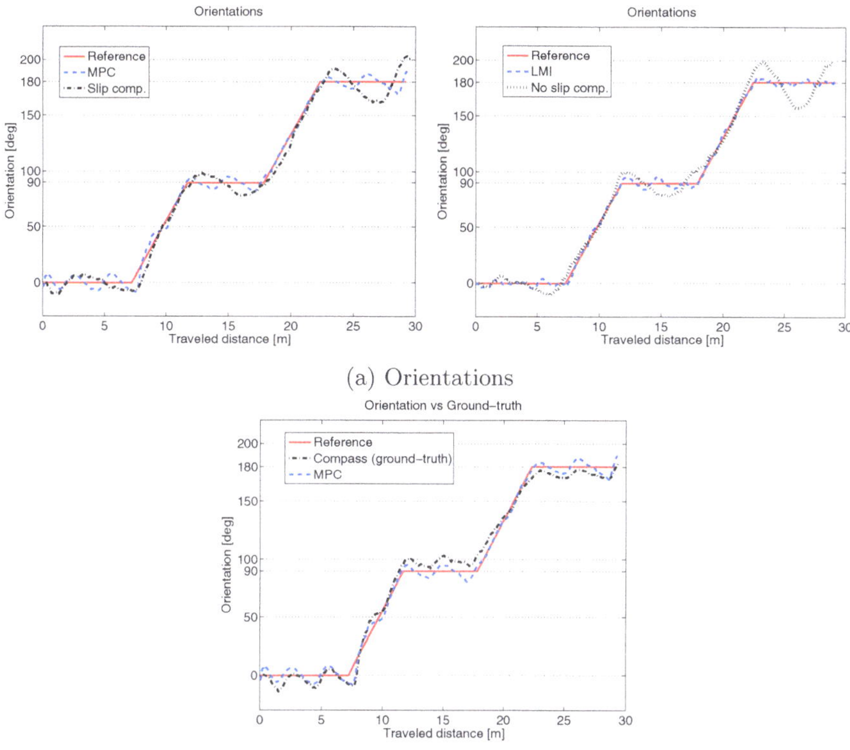

(b) MPC-orientation versus ground-truth (compass)

Fig. 5.9 Experiment 1. Orientations

As in previous chapter, both slip values are different. The reason is that the same actual velocity has been used for both tracks. In further works, two independent cameras will be employed to estimate the actual velocity of each track.

The error between the reference trajectory and those steered by the compared controllers are plotted in Figure 5.11. Especially remarkable is the almost zero lateral error achieved by the MPC controller. The mean longitudinal errors are: 0.02 [m] with standard deviation 0.03 [m] for the MPC controller, 0.02 [m] with standard deviation 0.03 [m] for the LMI-based controller, 0.09 [m] with standard deviation 0.05 [m] for the FL controller, and 0.13 [m] with standard deviation 0.06 [m] for the OC controller. The mean lateral errors are: 0 [m] with standard deviation 0.04 [m] for the predictive controller, -0.004 [m] with standard deviation 0.03 for the LMI-based controller, 0.02 [m] with standard deviation 0.17 [m] for the FL controller, and 0.01 [m] with standard deviation 0.18 [m] for the OC controller. The mean orientation errors are: 0.43 [deg] with standard deviation 4.74 [deg] for the

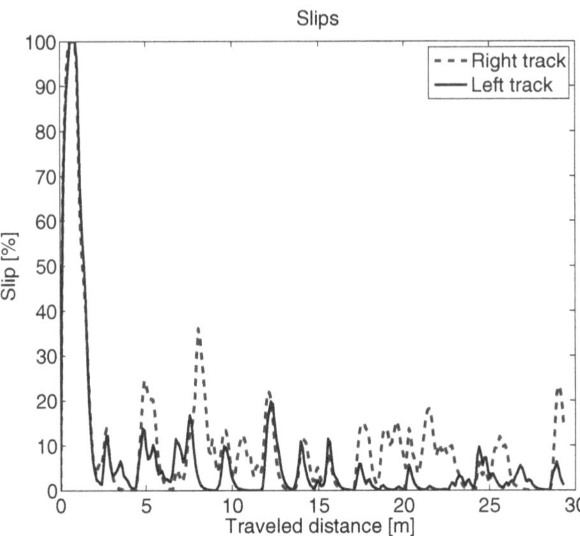

Fig. 5.10 Experiment 2. Slip estimations for the MPC controller (comparing the actual robot velocity obtained using visual odometry and track velocities obtained using encoders)

MPC controller, 0.26 [deg] with standard deviation 2.49 [deg] for the LMI-based controller, 0.66 [deg] with standard deviation 8.16 [deg] for the FL controller, and 1.09 [deg] with standard deviation 8.51 [deg] for the OC controller. As in the previous experiment, the elaborate control strategies obtain the best results.

Figure 5.12 displays the reference velocities, the control inputs, and the real track velocities for the case of the predictive controller. These plots notice the proper tuning of the PI controllers, since the real velocities of the tracks follow satisfactorily the set-points. It is important to clarify that the smooth oscillatory behaviour of the right track is stressed by that fact that the mobile robot was moving on a partially bumpy gravel soil. In this scenario, little stones produce vibrations to the vehicle, and hence to the tracks. Notice that in the previous chapter, a similar behaviour was observed for the LMI-based controller. In such experiment, the oscillatory behaviour was noticed for the left track (see Figure 4.13b).

Finally, Figure 5.13a displays the pixel displacement between consecutive images for the MPC controller (visual odometry). As in previous experiment, few outliers appear due to the proper downward camera position and the threshold filter. Notice that in this experiment the mean value of the pixel displacement is around -50 [pixel]. This confirms that the mobile robot was moving at a slower velocity than in previous experiment. Figure 5.13b shows

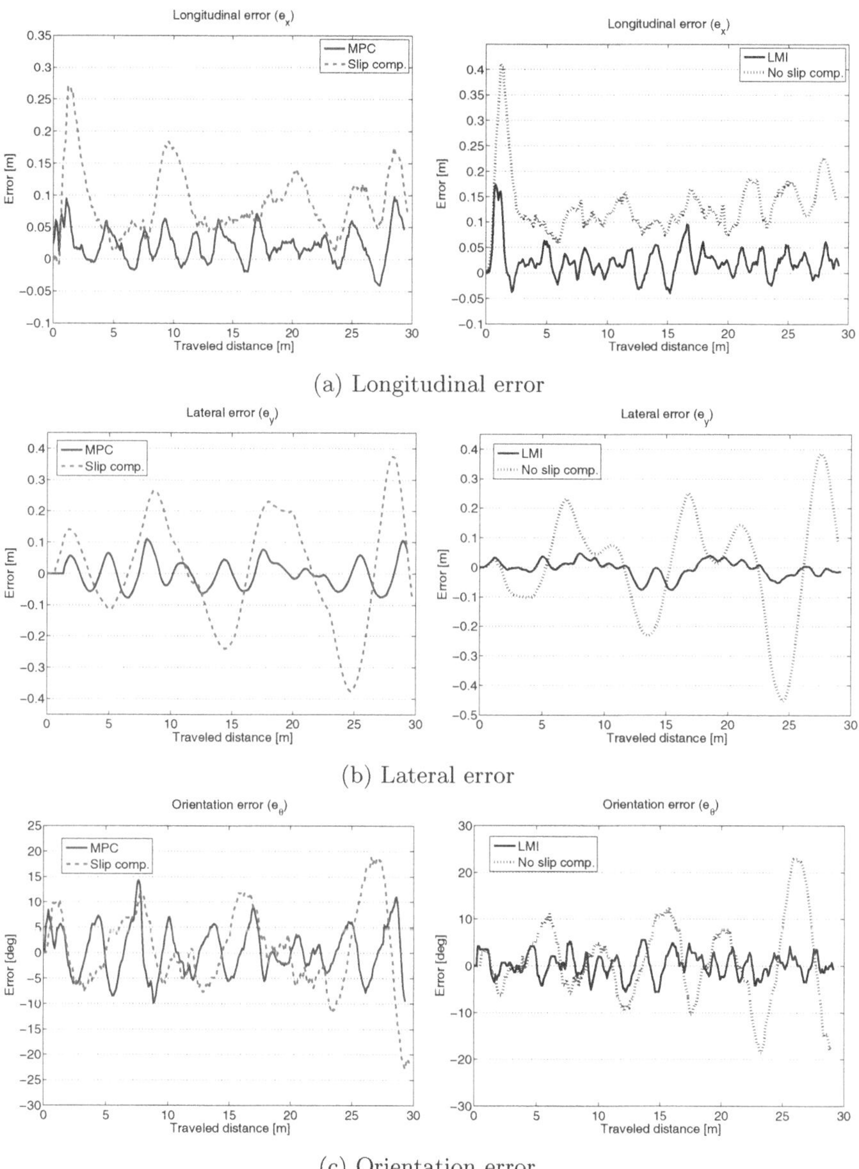

(a) Longitudinal error

(b) Lateral error

(c) Orientation error

Fig. 5.11 Experiment 2. Error along the travelled distance

the computation time employed by the predictive controller, the LMI-based controller, and the FL controller. The mean computation time for the predictive and the LMI-based controller is 0.24 [s] and for the FL controller is 0.19 [s].

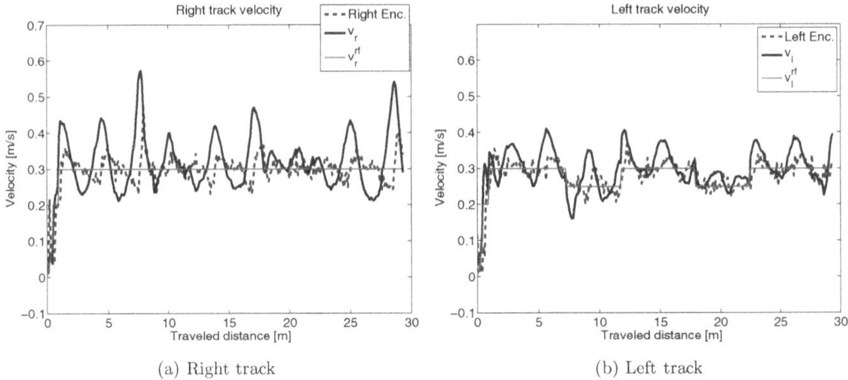

(a) Right track (b) Left track

Fig. 5.12 Experiment 2. Control signals and actual tracks velocities for the MPC controller

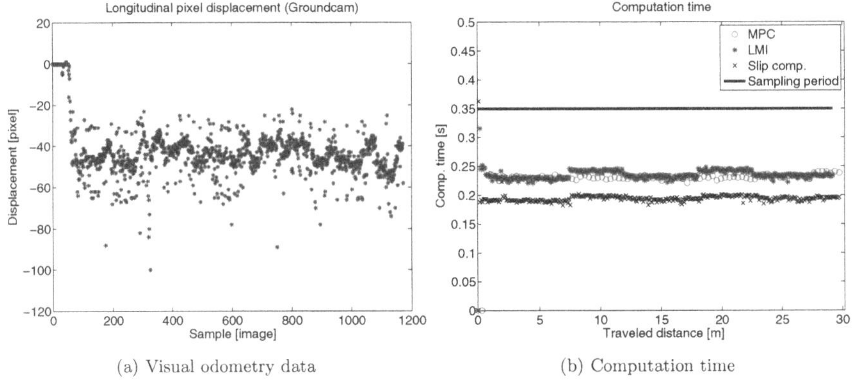

(a) Visual odometry data (b) Computation time

Fig. 5.13 Experiment 2. Pixel displacements (visual odometry) and computation time (CPU time employed at each sampling instant) for the MPC controller

5.5 Conclusions

The main contribution of this chapter is the successful adaptation and application of a robust tube-based MPC strategy to the mobile robotics field. In this case, the trajectory tracking error model of a mobile robot on slip conditions, previously discussed in Chapter 2, has been considered. The control objective is to steer an off-road mobile robot to the reference trajectory as close as possible at each sampling instant.

From the theoretical point of view, an extension of tube-based predictive control to time-varying systems is provided (via reachable sets). From a practical point of view, the designed control strategy compensates slip and guarantees that state and input constraints are fulfilled, since these constraints are taken into account in the solution of the optimization problem. The time-varying trajectory tracking error model has been addressed as a robust problem with additive uncertainties. Real-time execution has been ensured using tube-based MPC formulation and it has been tested through physical experiments. The comparative study with other control laws illustrates the excellent behaviour of the robust tube-based MPC strategy. Furthermore, the suggested robust tube-based predictive strategy runs efficiently on a limited performance computer (Intel Pentium III 1GHz, 512 MB RAM) at a computation time below 0.35 [s].

Chapter 6
Conclusions and Future Works

6.1 Conclusions

When a tracked mobile robot operates in off-road conditions many disturbances and inconveniences can lead to a unsuccessful result. Some of these inconveniences deal with robot-terrain interaction, such as slip and sinkage phenomena, and with inaccurate robot localization. In this context, both a careful and precise design and a strict testing process have to be carried out to achieve a satisfactory and reliable result. The most important topics to be analyzed are: mechanical robot configuration, robot modelling, path planning, motion control, and robot localization.

The objectives of this monograph were: to formulate several models for the trajectory tracking problem of tracked mobile robots taking into account slip effects; to provide a vision-based localization strategy; and the design of advanced slip compensation motion controllers ensuring constraints fulfillment and efficient real-time execution. To validate these contributions, a whole navigation architecture was designed for a real mobile robot. In this case, a tracked mobile robot, called *Fitorobot* and available at the University of Almería (Spain), was employed.

The main contributions of this monograph are summarized as follows.

First, a comprehensive review of the state of the art dealing with the particular problem of tracked locomotion has been addressed. Notice that we were focusing on the limited problem of tracked mobile robots in planar slippery conditions, then this book does not pretend to cover everything on off-road mobile robotics.

Secondly, an extended kinematic model considering slip effect was considered. This model avoids the estimation of complex variables usually related to dynamic models. Additionally, physical experiments have pointed out that the extended kinematic model is more accurate than the classic kinematic model when the mobile robot moves in off-road conditions. Afterwards, several formulations of a trajectory tracking error model were suggested using such extended kinematic model (continuous-time model, discrete-time model,

R. González, F. Rodríguez, and J.L. Guzmán, *Autonomous Tracked Robots in Planar Off-Road Conditions,* Studies in Systems, Decision and Control 6,
DOI: 10.1007/978-3-319-06038-5_6, © Springer International Publishing Switzerland 2014

and discrete-time model with additive uncertainties). These extended trajectory tracking error models were employed to design slip compensation motion controllers.

Regarding the localization issue, a visual-odometry-based technique was suggested. The most interesting point is that two cameras were combined, one for estimate the robot longitudinal displacement and another camera to estimate the robot orientation (visual compass). In this way, typical problems related to error growth of odometry-based solutions and false-matching phenomena of visual-odometry-based approaches are minimized. These localization techniques were validated through physical experiments. Those tests permitted to check the suitability of the proposed schemes in comparison with typical localization strategies such as wheel-based odometry.

Once the robot models and the localization techniques were developed and tested, several motion control strategies were presented. Firstly, the well-known linear feedback controller proposed in [21, 64] was modified. In this case, time-dependent feedback gains were updated online using the estimated slip. It is important to remark that, following the trajectory tracking paradigm, both the location and velocity of the robot are controlled. To difference of path tracking where only the steering motion of the robot is handled. Furthermore, this control strategy demands a really small computation load. The main drawback of this simple control strategy was that it did not take into account state and input constraints. For that reason, more elaborate control strategies were designed.

A slip compensation adaptive controller formulated using LMI was proposed. The procedure to obtain this adaptive controller was: first, a set of feedback gains were obtained offline, one gain for each extreme realization of the system dynamics. Then, an adaptive feedback control law was obtained online as a convex combination of the previously determined gains. The main advantages of this control strategy were that asymptotic stability was ensured, input and state constraints were fulfilled, and fast real-time execution was achieved. The main weakness of this control scheme was that uncertainty was not taken into account in the controller synthesis. In order to solve this issue, robust control techniques were analyzed.

A robust tube-based MPC controller was also suggested. The main advantages of this control strategy were: robustness (uncertainties were taken into account in the controller synthesis), and input and state constraints fulfillment. From the theoretical point of view, it was provided the extension of tube-based predictive control to time-varying systems (via reachable sets). From the practical point of view, a robust predictive control strategy running efficiently in real-time was successfully applied to steer an off-road mobile robot.

Finally, a whole navigation architecture was designed to test closed-loop physical experiments using a tracked mobile robot. In this way, motion control strategies were validated in real off-road conditions. The obtained results demonstrate the satisfactory behaviour of slip compensation controllers in comparison with controllers with no slip compensation and the proper integration of the visual-odometry-based localization technique. Notice that this navigation architecture can be easily adapted to any kind of mobile robots.

6.2 Future Works

From the modelling point of view, it would be interesting to extend the current trajectory tracking error model based on robot kinematics taking into account dynamic effects (tractive forces, robot mass, inertia, etc.). This work will require new models dealing with terra-mechanics (terrain classification and characterisation) and to measure online some parameters related to track-soil interactions. For that reason, new sensors should be installed on board the robot. Furthermore, motion controllers will be reformulated taking into account these dynamic variables, what will lead to new sets of constraints in terms of maximization of robot traction and minimization of slip.

From the localization point of view, an attractive future work will consist in improving the main weakness of the presented vision-based scheme, that is, the false matches due to shadows. In this way, a probabilistic localization scheme can be employed combining visual odometry with other localization strategies. Minor improvements can be to analyze new positions for the cameras and new filters or algorithms to reduce the effect of false matches.

Another future work can be to extended the current robust predictive control scheme updating online the uncertainty set using the information given by a Kalman filter about the uncertainty on the robot location. As known, the Kalman filter strategy generates at each sampling instant a covariance matrix, which comprises the uncertainty on the robot location in terms of a Gaussian probability distribution. The challenge about this issue is that two different data structures are involved, the covariance matrix (statistical data) and the additive uncertainty term (polytope). A first possible solution is via the mathematical structures called "Gaussian random polytopes" [5]. This approach is based on generating n random points with respect to a uniform distribution and obtaining the convex hull of these points, which leads to polytopic structures. Another possible solution is considering the stochastic MPC framework [70]. Here, states are regarded as stochastic variables and instead of minimizing a deterministic objective function, the expected value of a objective function is minimized.

Finally, a degree of reactiveness will be added to the current navigation architecture permitting to the mobile robot to circumnavigate obstacles and to avoid unpredicted risky situations. This improvement will enable to meet a wider area of application of the work presented in this monograph.

References

[1] Ahmadi, M., Polotski, V., Hurteau, R.: Path Tracking Control of Tracked Vehicles. In: IEEE International Conference on Robotics and Automation, San Francisco, USA, vol. 1, pp. 2938–2943. IEEE (April 2000)

[2] Angelova, A., Matthies, L., Helmick, D., Perona, P.: Learning and Prediction of Slip from Visual Information. Journal of Field Robotics 24(3), 205–231 (2007)

[3] Armingol, J.M., de la Escalera, A., Moreno, L., Salichs, M.A.: Mobile Robot Localization using a Non-linear Evolutionary Filter. Advanced Robotics 16(7), 629–652 (2002)

[4] Aström, K.J., Murray, R.M.: Feedback Systems: An Introduction for Scientists and Engineers. Princeton University Press, USA (2008)

[5] Barany, I., Vu, V.: Central Limit Theorems for Gaussian Polytopes. Annals of Probability 35(4), 1593–1621 (2007)

[6] Bekker, M.G.: Theory of Land Locomotion. The Mechanics of Vehicle Mobility, 1st edn. The Univiersity of Michigan Press, Ann Arbor (1956)

[7] Bemporad, A., Borrelli, F., Morari, M.: Model Predictive Control Based on Linear Programming - The Explicit Solution. IEEE Transactions on Automatic Control 47(12), 1974–1985 (2002)

[8] Bemporad, A., Morari, M.: Robust Model Predictive Control: A Survey. In: Garulli, A., Tesi, A., Vicino, A. (eds.) Robustness in Identification and Control. LNCIS, vol. 245, pp. 207–226. Springer, Heidelberg (1999)

[9] Benoit, O., Gotteland, P., Quibel, A.: Prediction of Trafficability for Tracked Vehicle on Broken Soil: Real Size Tests. Journal of Terramechanics 40(2), 135–160 (2003)

[10] Bertsekas, D.P., Rhodes, I.B.: On the Minimax Reachability of Target Sets and Target Tubes. Automatica 7, 233–247 (1971)

[11] Bigras, P., Petrov, P., Wong, T.: A LMI Approach to Feedback Path Control for an Articulated Mining Vehicle. In: International Conference on Modeling and Simulation of Electric Machines, Converters and Systems, IMACS, Montreal, Canada (August 2002)

[12] Blanchini, F.: Set Invariance in Control. Automatica 35(11), 1747–1767 (1999)

[13] Blanchini, F., Miani, S.: Set–Theoretic Methods in Control, 1st edn. Birkhauser, USA (2007)

[14] Borenstein, J.: The CLAPPER: A Dual–drive Mobile Robot with Internal Correction of Dead–reckoning Errors. In: IEEE Conference on Robotics and Automation, San Diego, USA, pp. 3085–3090. IEEE (May 1994)
[15] Boyd, S., El Ghaoui, L., Feron, E., Balakrishnan, V.: Linear Matrix Inequalities in System and Control Theory. Society for Industrial and Applied Mathematics, USA (1994)
[16] Bradski, G., Kaehler, A.: Learning OpenCV: Computer Vision with the OpenCV Library. O'Reilly Ed., USA (2008)
[17] Brunelli, R.: Template Matching Techniques in Computer Vision: Theory and Practice. John Wiley & Sons, USA (2009)
[18] Camacho, E.F., Bordons, C.: Model Predictive Control, 2nd edn. Advanced Textbooks in Control and Signal Processing. Springer, Germany (2004)
[19] Campbell, J., Sukthankar, R., Nourbakhsh, I., Pahwa, A.: A Robust Visual Odometry and Precipice Detection System Using Consumer–grade Monocular Vision. In: IEEE International Conference on Robotics and Automation, Barcelona, Spain, pp. 3421–3427. IEEE (April 2005)
[20] Campion, G., Bastin, G., D' Andréa-Novel, B.: Structural Properties and Classification of Kinematic and Dynamic Models of Wheeled Mobile Robots. IEEE Transactions on Robotics and Automation 12(1), 47–62 (1996)
[21] Canudas, C., Siciliano, B., Bastin, G.: Theory of Robot Control, 2nd edn. Communications and Control Engineering. Springer, Germany (1997)
[22] Cariou, C., Lenain, R., Thuilot, B., Berducat, M.: Automatic Guidance of a Four-Wheel-Steering Mobile Robot for Accurate Field Operations. Journal of Field Robotics 26(6-7), 504–518 (2009)
[23] Cheng, Y., Maimone, M., Matthies, L.: Visual Odometry on the Mars Exploration Rovers. In: IEEE International Conference on Systems, Man and Cybernetics, Waikoloa, USA, pp. 903–910. IEEE (October 2005)
[24] Chisci, L., Rossiter, J.A., Zappa, G.: Systems with Persistent Disturbances: Predictive Control with Restricted Constraints. Automatica 37(7), 1019–1028 (2001)
[25] Corradini, M.L., Orlando, G.: Robust Tracking Control of Mobile Robots in the Presence of Uncertainties in the Dynamical Model. Journal of Robotic Systems 18(6), 317–323 (2001)
[26] Corradini, M.L., Orlando, G.: Control of Mobile Robots with Uncertainties in the Dynamical Model: A Discrete Time Sliding Mode Approach with Experimental Results. Control Engineering Practice 10(1), 23–34 (2002)
[27] Dille, M., Grocholsky, B.P., Singh, S.: Outdoor Downward-facing Optical Flow Odometry with Commodity Sensors. In: International Conference on Field and Service Robotics, Cambridge, USA, pp. 183–193 (July 2009)
[28] Dong, W., Kuhnert, K.D.: Robust Adaptive Control of Nonholonomic Mobile Robot with Parameter and Nonparameter Uncertainties. IEEE Transactions on Robotics 21(2), 261–266 (2005)
[29] Durrant-Whyte, H., Bailey, T.: Simultaneous Localisation and Mapping (SLAM). IEEE Robotics and Automation Magazine 13(2), 99–110 (2006)
[30] Durrant-Whyte, H., Leonard, J.: Mobile Robot Localization by Tracking Geometric Beacons. IEEE Transactions on Robotics and Automation 7(3), 376–382 (1991)
[31] Endo, D., Okada, Y., Nagatani, K., Yoshida, K.: Path Following Control for Tracked Vehicles Based on Slip–Compensating Odometry. In: IEEE International Conference on Intelligent Robots and Systems, San Diego, USA, pp. 2871–2876. IEEE (October 2007)

[32] Fan, Z., Koren, Y., Wehe, D.: Tracked Mobile Robot Control: Hybrid Approach. Control Engineering Practice 3(3), 329–336 (1995)

[33] Fang, H., Fan, R., Thuilot, B., Martinet, P.: Trajectory Tracking Control of Farm Vehicles in Presence of Sliding. Robotics and Autonomous Systems 54(10), 828–839 (2006)

[34] Fiacchini, M.: Convex Difference Inclusions for Systems Analysis and Design. PhD Thesis, University of Seville, Seville, Spain (2010)

[35] Gahinet, P., Nemirovski, A., Laub, A.J., Chilali, M.: LMI Control Toolbox User's Guide, 1st edn. The MathWorks, Inc., USA (1995)

[36] Garber, M., Wong, J.Y.: Prediction of ground pressure distribution under tracked vehicles. Journal of Terramechanics 18(1-2) (1981)

[37] Gilbert, E.G., Tan, K.T.: Linear Systems with State and Control Constraints: The Theory and Application of Maximal Output Admissible Sets. IEEE Transactions on Automatic Control 36(9), 1008–1020 (1991)

[38] Gómez, J., Camacho, E.F.: Mobile Robot Navigation in a Partially Structured Static Environment, Using Neural Predictive Control. Control Engineering Practice 4(12), 1669–1679 (1996)

[39] González, R.: Localization of the CRAB rover using Visual Odometry. Tech. report, Autonomous Systems Lab, ETH Zürich (Switzerland) (December 2009), `http://www.ual.es/personal/rgonzalez/english/publications.htm`

[40] González, R., Fiacchini, M., Álamo, T., Guzmán, J.L., Rodríguez, F.: Adaptive Control for a Mobile Robot under Slip Conditions using an LMI-based Approach. European Journal of Control 16(2), 144–155 (2010)

[41] González, R., Fiacchini, M., Guzmán, J.L., Álamo, T., Rodríguez, F.: Robust Tube-based Predictive Control for Mobile Robots in Off-Road Conditions. Robotics and Autonomous Systems 59(10), 711–726 (2011)

[42] González, R., Rodríguez, F., Guzmán, J.L., Pradalier, C., Siegwart, R.: Control of off-road mobile robots using visual odometry and slip compensation. Advanced Robotics 27(11), 893–906 (2013)

[43] González, R., Rodríguez, F., Sánchez-Hermosilla, J., Donaire, J.G.: Navigation Techniques for Mobile Robots in Greenhouses. Applied Engineering in Agriculture 25(2), 153–165 (2009)

[44] Goshtasby, A., Gage, S.H., Bartholic, J.F.: A Two–Stage Correlation Approach to Template Matching. IEEE Transactions on Pattern Analysis and Machine Intelligence 6(3), 374–378 (1984)

[45] Goulart, P.J., Kerrigan, E.C., Maciejowski, J.M.: Optimization over State Feedback Policies for Robust Control with Constraints. Automatica 42(4), 523–533 (2006)

[46] Gracia, L., Tornero, J.: Kinematic Modeling of Wheeled Mobile Robots with Slip. Advanced Robotics 21(11), 1253–1279 (2007)

[47] Guarnieri, M., Debenest, P., Inoh, T., Fukushima, E., Hirose, S.: Development of helios vii: an arm-equipped tracked vehicle for search and rescue operations. In: IEEE Int. Conf. on Intelligent Robots and Systems (IROS), pp. 39–45. IEEE (2004)

[48] Guzmán, J.L., Berenguel, M., Rodríguez, F., Dormido, S.: An Interactive Tool for Mobile Robot Motion Planning. Robotics and Autonomous Systems 56(5), 396–409 (2008)

[49] Guzmán, J.L., Álamo, T., Berenguel, M., Dormido, S., Camacho, E.F.: A Robust Constrained Reference Governor Approach using Linear Matrix Inequalities. Journal of Process Control 19(5), 773–784 (2009)

[50] Helmick, D., Roumeliotis, S., Cheng, Y., Clouse, D.S., Bajracharya, M., Matthies, L.H.: Slip–compensated Path Following for Planetary Exploration Rovers. Advanced Robotics 20(11), 1257–1280 (2006)

[51] Hofmann-Wellenhof, B., Lichtenegger, H., Collins, J.: Global Positioning System: Theory and Practice, 5th edn. Springer, Germany (2001)

[52] Hohl, G.H.: Military terrain vehicles. Journal of Terramechanics 44, 23–34 (2007)

[53] Höpflinger, M., Krebs, A., Pradalier, C., Lee, C., Obstei, R., Siegwart, R.: Description of the Locomotion Control Architecture on the ExoMars Rover Breadboard. In: ESA Workshop on Advanced Space Technologies for Robotics and Automation, Noordwijk, The Netherlands (November 2008)

[54] Hornback, P.: The Wheel versus Track Dilemma. Armor Magazine 107(2), 33–34 (1998)

[55] Hornung, A., Bennewitz, M., Strasdat, H.: Efficient Vision–based Navigation. Learning about the Influence of Motion Blur. Autonomous Robots 29(2), 137–149 (2010)

[56] Horowitz, I.M.: Synthesis of Feedback Systems with Nonlinear Time-Varying Uncertain Plants to Satisfy Quantitative Performance Specifications. Proceedings of the IEEE 64(1), 123–130 (1976)

[57] Iagnemma, K., Dubowsky, S.: Mobile Robot Rough–Terrain Control (RTC) for Planetary Exploration. In: ASME Biennial Mechanisms and Robotics Conference, Baltimore, USA, pp. 10–13. ASME (September 2000)

[58] Iagnemma, K., Dubowsky, S.: Mobile Robots in Rough Terrain. Estimation, Motion Planning, and Control with Application to Planetary Rovers. Springer Tracts in Advanced Robotics. Springer, Germany (2004)

[59] Iagnemma, K., Kang, S., Shibly, H., Dubowsky, S.: Online Terrain Parameter Estimation for Wheeled Mobile Robots with Application to Planetary Rovers. IEEE Transactions on Robotics 20(5), 921–927 (2004)

[60] Iagnemma, K., Ward, C.C.: Classification–based Wheel Slip Detection and Detector Fusion for Mobile Robots on Outdoor Terrain. Autonomous Robots 26(1), 33–46 (2009)

[61] Ishigami, G., Nagatani, K., Yoshida, K.: Slope Traversal Experiments with Slip Compensation Control for Lunar/Planetary Exploration Rover. In: IEEE International Conference on Robotics and Automation, Pasadena, USA, pp. 2295–2300. IEEE (May 2008)

[62] Ishigami, G., Nagatani, K., Yoshida, K.: Slope Traversal Controls for Planetary Exploration Rover on Sandy Terrain. Journal of Field Robotics 26(3), 264–286 (2009)

[63] Johnson, A.E., Goldberg, S.B., Yang, C., Matthies, L.H.: Robust and Efficient Stereo Feature Tracking for Visual Odometry. In: IEEE International Conference on Robotics and Automation, Pasadena, USA, pp. 39–46. IEEE (May 2008)

[64] Kanayama, Y., Kimura, Y., Miyazaki, F., Noguchi, T.: A Stable Tracking Control Method for an Autonomous Mobile Robots. In: IEEE International Conference on Robotics and Automation, Cincinnati, USA, pp. 384–389. IEEE (1990)

[65] Kim, S., Lee, S.: Robust Mobile Robot Velocity Estimation using a Polygonal Array of Optical Mice. International Journal of Information Acquisition 5(4), 321–330 (2008)

[66] Kitano, M., Kuma, M.: An Analysis of Horizontal Plane Motion of Tracked Vehicles. Journal of Terramechanics 14(4), 211–225 (1977)
[67] Klancar, G., Skrjanc, I.: Tracking–error Model–based Predictive Control for Mobile Robots in Real Time. Robotics and Autonomous Systems 55(1), 460–469 (2007)
[68] Kolmanovsky, I., Gilbert, E.G.: Theory and Computation of Disturbance Invariant Sets for Discrete–time Linear Systems. Mathematical Problems in Engineering 4(4), 317–367 (1998)
[69] Kothare, M.V., Balakrishnan, V., Morari, M.: Robust Constrained Model Predictive Control using Linear Matrix Inequalities. Automatica 32(10), 1361–1379 (1996)
[70] Kouvaritakis, B., Cannon, M., Tsachouridis, V.: Recent Developments in Stochastic MPC and Sustainable Development. Annual Reviews in Control 28(1), 23–35 (2004)
[71] Krebs, A., Thueer, T., Carrasco, E., Siegwart, R.: Towards Torque Control of the CRAB Rover. In: International Symposium on Artificial Intelligence, Robotics and Automation in Space, Los Angeles, USA (February 2008)
[72] Kvasnica, M., Grieder, P., Baotić, M.: Multi–Parametric Toolbox, MPT (2004), http://control.ee.ethz.ch/~mpt/
[73] Labrosse, F.: The Visual Compass: Performance and Limitations of an Appearance–Based Method. Journal of Field Robotics 23(10), 913–941 (2006)
[74] Lamon, P., Siegwart, R.: 3D Position Tracking in Challenging Terrain. The International Journal of Robotics Research 26(2), 167–186 (2007)
[75] Langson, W., Chryssochoos, I., Rakovic, S.V., Mayne, D.Q.: Robust Model Predictive Control using Tubes. Automatica 40(1), 125–133 (2004)
[76] Le, A.T.: Modelling and Control of Tracked Vehicles. PhD Thesis, University of Sydney, Sydney, Australia (1999)
[77] Le, A.T., Rye, D.C., Durrant-Whyte, H.F.: Estimation of Track–soil Interactions for Autonomous Tracked Vehicles. In: IEEE International Conference on Robotics and Automation, Albuquerque, USA, pp. 1388–1393. IEEE (April 1997)
[78] Lenain, R., Thuilot, B., Cariou, C., Martinet, P.: Adaptive and Predictive Path Tracking Control for Off–road Mobile Robots. European Journal of Control 13(4), 419–439 (2007)
[79] Lenain, R., Thuilot, B., Cariou, C., Martinet, P.: Mixed Kinematic and Dynamic Sideslip Angle Observer for Accurate Control of Fast Off–Road Mobile Robots. Journal of Field Robotics 27(2), 181–196 (2010)
[80] Lenain, R., Thuilot, B., Cariu, C., Martinet, P.: Model Predictive Control for Vehicle Guidance in Presence of Sliding: Application to Farm Vehicles Path Tracking. In: IEEE International Conference on Robotics and Automation, Barcelona, Spain, pp. 885–890. IEEE (April 2005)
[81] Li, Q., Ayers, P.D., Anderson, A.B.: Modeling of terrain impact caused by tracked vehicles. Jornal of Terramechanics 44(6), 395–410 (2007)
[82] Limón, D., Álamo, T., Camacho, E.F.: Input-to-state Stable MPC for Constrained Discrete-time Nonlinear Systems with Bounded Additive Uncertainties. In: IEEE Conference on Decision and Control, Las Vegas, USA, pp. 4619–4624. IEEE (December 2002)
[83] Limón, D., Alvarado, I., Álamo, T., Camacho, E.F.: Robust Tube–based MPC for Tracking of Constrained Linear Systems with Additive Disturbances. Journal of Process Control 20(3), 248–260 (2010)

[84] Limón, D., Bravo, J.M., Álamo, T., Camacho, E.F.: Robust MPC of Constrained Nonlinear Systems Based on Interval Arithmetic. IET Control Theory & Applications 152(3), 325–332 (2005)

[85] Lowe, D.G.: Distinctive Image Features from Scale–Invariant Keypoints. International Journal of Computer Vision 60(2), 91–110 (2004)

[86] Lucas, B.D., Kanade, T.: An Iterative Image Registration Technique with an Application to Stereo Vision. In: DARPA Imaging Understanding Workshop, DARPA, Monterey, USA, pp. 121–130 (April 1981)

[87] Mann, M., Shiller, Z.: Dynamic Stability of Off–road Vehicles – Quasi–3D Analysis. In: IEEE International Conference on Robotics and Automation, Pasadena, USA, pp. 2301–2306. IEEE (May 2008)

[88] Martínez, J.L., Mandow, A., Morales, J., Pedraza, S., García-Cerezo, A.: Approximating Kinematics for Tracked Mobile Robots. The International Journal of Robotics Research 24(10), 867–878 (2005)

[89] Masehian, E., Habibi, G.: Motion Planning and Control of Mobile Robot using Linear Matrix Inequalities (LMIs). In: IEEE International Conference on Intelligent Robots and Systems, San Diego, USA, pp. 4277–4282. IEEE (October 2007)

[90] Matiukhin, V.: Trajectory Stabilization of Wheeled System. In: IFAC World Congress, IFAC, Seoul, Korea, pp. 1177–1182 (2008)

[91] Matthies, L.: Dynamic Stereo Vision. PhD Thesis, Carnegie Mellon University, Pittsburgh, USA (1989)

[92] Matthies, L., Maimone, M., Johnson, A., Cheng, Y., Willson, R., Villalpando, C., Goldberg, S., Huertas, A.: Computer Vision on Mars. International Journal of Computer Vision 75(1), 67–92 (2007)

[93] Maybeck, P.S.: Stochastic Models, Estimation, and Control. Mathematics in Science and Engineering. Academic Press, Inc., USA (1979)

[94] Mayne, D.Q., Langson, W.: Robustifying Model Predictive Control of Constrained Linear Systems. Electronics Letters 37(23), 1422–1423 (2001)

[95] Mayne, D.Q., Rakovic, S.V., Findeisen, R., Allgower, F.: Robust Output Feedback Model Predictive Control of Constrained Linear Systems: Time Varying Case. Automatica 45(9), 2082–2087 (2009)

[96] Mayne, D.Q., Rawlings, J.B., Rao, C.V., Scokaert, P.O.: Constrained Model Predictive Control: Stability and Optimality. Automatica 36(6), 789–814 (2000)

[97] McNae, A.: A history of komatsu: Construction and mining equipment, Beenleigh, Qld. (2000)

[98] Morales, J., Martinez, J.L., Mandow, A., Garcia-Cerezo, A.J., Pedraza, S.: Power Consumption Modeling of Skid–Steer Tracked Mobile Robots on Rigid Terrain. IEEE Transactions on Robotics 25(5), 1098–1108 (2009)

[99] Morari, M., Lee, J.H.: Model Predictive Control: Past, Present and Future. Computers & Chemical Engineering 23(4), 667–682 (1999)

[100] Mourikis, A.I., Trawny, N., Roumeliotis, S., Helmick, D., Matthies, L.: Autonomous stair climbing for tracked vehicles. International Journal of Robotics Research 26(7), 737–758 (2007)

[101] Nagatani, K., Ikeda, A., Ishigami, G., Yoshida, K., Nagai, I.: Development of a Visual Odometry System for a Wheeled Robot on Loose Soil using a Telecentric Camera. Advanced Robotics 24(8–9), 1149–1167 (2010)

[102] Nistér, D., Naroditsky, O., Bergen, J.: Visual Odometry for Ground Vehicle Applications. Journal of Field Robotics 23(1), 3–20 (2006)

[103] Normey-Rico, J.E., Gómez-Ortega, J., Camacho, E.F.: A Smith–predictor–based Generalised Predictive Controller for Mobile Robot Path–tracking. Control Engineering Practice 7(6), 729–740 (1999)

[104] Nourani-Vatani, N., Roberts, J., Srinivasan, M.V.: Practical Visual Odometry for Car–like Vehicles. In: IEEE International Conference on Robotics and Automation, Kobe, Japan, pp. 3551–3557. IEEE (May 2009)

[105] Olson, C.F., Matthies, L.H., Schoppers, M., Maimone, M.W.: Rover Navigation using Stereo Ego–motion. Robotics and Autonomous Systems 43(4), 215–229 (2003)

[106] Oriolo, G., De Luca, A., Vendittelli, M.: WMR Control Via Dynamic Feedback Lineralization: Design, Implementation, and Experimental Validation. IEEE Transactions on Control Systems Technology 10(6), 835–852 (2002)

[107] Packard, A.K., Doyle, J.C., Balas, G.J.: Linear, Multivariable Robust Control with a μ Perspective. ASME Journal of Dynamics, Measurements and Control 115(2), 426–438 (1993)

[108] Park, W.Y., Chang, Y.C., Lee, S.S., Hong, J.H., Park, J.G., Lee, K.S.: Prediction of the tractive performance of a flexible tracked vehicle. Jornal of Terramechanics 45(1-2), 13–23 (2008)

[109] Pourboghrat, F., Karlsson, M.P.: Adaptive Control of Dynamic Mobile Robots with Nonholonomic Constraints. Computers and Electrical Engineering 28(4), 241–253 (2002)

[110] Pretto, A., Menegatti, E., Bennewitz, M., Burgard, W., Pagello, E.: A Visual Odometry Framework Robust to Motion Blur. In: IEEE International Conference on Robotics and Automation, Kobe, Japan, pp. 1685–1692. IEEE (May 2009)

[111] Rawlings, J.B., Mayne, D.Q.: Model Predictive Control: Theory and design. Nob Hill Publishing, LLC (2009)

[112] Reina, G., Ishigami, G., Nagatani, K., Yoshida, K.: Vision–based Estimation of Slip Angle for Mobile Robots and Planetary Rovers. In: IEEE International Conference on Robotics and Automation, Pasadena, USA, pp. 486–491. IEEE (May 2008)

[113] Rodgers, J.L., Nicewander, W.A.: Thirteen Ways to Look at the Correlation Coefficient. The American Statistician 42(1), 59–66 (1988)

[114] Sánchez-Gimeno, A., Sánchez-Hermosilla, J., Rodríguez, F., Berenguel, M., Guzmán, J.L.: Self–Propelled Vehicle for Agricultural Tasks in Greenhouses. Agricultural Engineering for a Better World, European Society of Agricultural Engineers, Bonn, Germany (September 2006)

[115] Scaramuzza, D.: Omnidirectional Vision: From Calibration to Robot Motion Estimation. PhD Thesis, Swiss Federal Institute of Technology, Zürich, Switzerland (2008)

[116] Scheding, S., Dissanayake, G., Nebot, E., Durrant-Whyte, H.: Slip Modelling and Aided Inertial Navigation of an LHD. In: IEEE International Conference on Robotics and Automation, Albuquerque, USA, pp. 1904–1909. IEEE (April 1997)

[117] Siegwart, R., Lamon, P., Estier, T., Lauria, M., Piguet, R.: Innovative Design for Wheeled Locomotion in Rough Terrain. Robotics and Autonomous Systems 40(2-3), 151–162 (2002)

[118] Siegwart, R., Nourbakhsh, I.: Introduction to Autonomous Mobile Robots, 1st edn. A Bradford book. The MIT Press, USA (2004)

[119] Srinivasan, M.V.: An Image–interpolation Technique for the Computation of Optic Flow and Egomotion. Journal of Biological Cybernetics 71(5), 401–415 (1994)

[120] Sturm, J., Visser, A.: An Appearance–based Visual Compass for Mobile Robots. Robotics and Autonomous Systems 57(5), 536–545 (2009)

[121] Tanaka, K., Kosaki, T., Wang, H.O.: Backing Control Problem of a Mobile Robot with Multiple Trailers: Fuzzy Modeling and LMI–Based Design. IEEE Transactions on Systems, Man and Cybernetics 28(3), 329–337 (1998)

[122] Tardos, J.D., Neira, J., Newman, P.M., Leonard, J.J.: Robust Mapping and Localization in Indoor Environments Using Sonar Data. The International Journal of Robotics Research 21(4), 311–330 (2002)

[123] Thrun, S., Burgard, W., Fox, D.: Probabilistic Robotics. Intelligent Robotics and Autonomous Agents Series. The MIT Press, USA (2005)

[124] Thrun, S., Thayer, S., Whittaker, W., Baker, C., Burgard, W., Ferguson, D., Haehnel, D., Montemerlo, M., Morris, A.C., Omohundro, Z., Reverte, C., Whittaker, W.L.: Autonomous Exploration and Mapping of Abandoned Mines. IEEE Robotics and Automation Magazine 11(1), 79–91 (2004)

[125] Thueer, T., Krebs, A., Siegwart, R.: Performance Comparison of Rough–Terrain Robots – Simulation and Hardware. Journal of Field Robotics 24(3), 251–271 (2007)

[126] Trodden, P., Richards, A.: Distributed Model Predictive Control of Linear Systems with Persistent Disturbances. International Journal of Control 83(8), 1653–1663 (2010)

[127] Vinh, T.Q., Giap, N.H., Kim, T.W., Jeong, M.G., Shin, J.H., Kim, W.H.: Adaptive Robust Fuzzy Control and Implementation for Path Tracking of a Mobile Robot. In: IEEE International Symposium on Industrial Electronics, Seoul, Korea, pp. 1943–1949. IEEE (July 2009)

[128] Wan, J., Vehi, J., Luo, N.: A Numerical Approach to Design Control Invariant Sets for Constrained Nonlinear Discrete–time Systems with Guaranteed Optimality. Journal of Global Optimization 44(3), 395–407 (2009)

[129] Wang, D., Low, C.B.: Modeling and Analysis of Skidding and Slipping in Wheeled Mobile Robots: Control Design Perspective. IEEE Transactions on Robotics 24(3), 676–687 (2008)

[130] Ward, C.C., Iagnemma, K.: A Dynamic–Model–Based Wheel Slip Detector for Mobile Robots on Outdoor Terrain. IEEE Transactions on Robotics 24(4), 821–831 (2008)

[131] Welch, G., Bishop, G.: An Introduction to the Kalman Filter. In: SIGGRAPH 2001 Course, Los Angeles, USA. ACM Press, Addison–Wesley (August 2001)

[132] Wong, J.Y.: Theory of Ground Vehicles, 3rd edn. John Wiley & Sons, Inc., USA (2001)

[133] Wong, J.Y., Huang, W.: Wheels vs. Tracks – A Fundamental Evaluation From the Traction Perspective. Journal of Terramechanics 43(1), 27–42 (2006)

[134] Yang, T., Liu, Z., Chen, H., Pei, R.: Distributed Robust Control of Multiple Mobile Robots Formations via Moving Horizon Strategy. In: American Control Conference, Minneapolis, USA, pp. 2838–2843. IEEE (June 2006)

[135] Yang, T.T., Liu, Z.Y., Chen, H., Pei, R.: The Research on Robust Tracking Control of Constrained Wheeled Mobile Robots. In: International Conference on Machine Learning and Cybernetics, Guangzhou, Japan, pp. 1356–1361. IEEE (August 2005)

[136] Yi, J., Song, D., Zhang, J., Goodwin, Z.: Adaptive Trajectory Tracking Control of Skid–Steered Mobile Robots. In: International Conference on Robotics and Automation, Rome, Italy, pp. 2605–2610. IEEE (April 2007)

[137] Yi, J., Wang, H., Zhang, J., Song, D., Jayasuriya, S., Liu, J.: Kinematic Modeling and Analysis of Skid–Steered Mobile Robots With Applications to Low–Cost Inertial–Measurement–Unit Based Motion Estimation. IEEE Transactions on Robotics 25(5), 1087–1097 (2009)

[138] Zames, G., Francis, B.A.: Feedback, Minimax Sensitivity, and Optimal Robustness. IEEE Transactions on Automatic Control 28(5), 585–601 (1983)

Zeitfracht Medien GmbH
Ferdinand-Jühlke-Straße 7
99095 Erfurt, Deutschland
produktsicherheit@kolibri360.de